D0939075

Cosmology Revealed: Living Inside the Cosmic Egg

Springer
London
Berlin
Heidelberg
New York
Barcelona
Hong Kong
Milan
Paris
Santa Clara
Singapore
Tokyo

Anthony Fairall

Cosmology Revealed: Living Inside the Cosmic Egg

Professor Anthony Fairall
Department of Astronomy
University of Cape Town
South Africa

Planetarium
South African Museum
Cape Town

SPRINGER–PRAXIS BOOKS IN POPULAR ASTRONOMY
SUBJECT *ADVISORY EDITOR*: Stuart Clark, B.Sc., Ph.D.
PROJECT EDITOR: Fiona Gold

ISBN 1-85233-322-7 Springer-Verlag Berlin Heidelberg New York

British Library Cataloguing-in-Publication Data
Fairall, Anthony
Cosmology revealed: living inside the cosmic egg. –
(Springer-Praxis books in popular astronomy)
1. Cosmology
I. Title II. Praxis Publishing
523.1

ISBN 1-85233-322-7

Library of Congress Cataloging-in-Publication Data
Fairall, Anthony.
Cosmology revealed: living inside the cosmic egg / Anthony Fairall.
p. cm. – (Springer-Praxis books in popular astronomy)
Includes index.
ISBN 1-85233-322-7 (alk. paper)
1. Cosmology–Popular works. I. Title: Living inside the cosmic egg. II. Title. III.
Springer-Praxis series in astronomy and space sciences.

QB982.F35 2000
523.1–dc21 00-063575

Printed by MPG Books Ltd, Bodmin, Cornwall, UK

Cover design: Jim Wilkie
Typesetting: BookEns Ltd, Royston, Herts., UK

Printed on acid-free paper supplied by Precision Publishing Papers Ltd, UK

To Alex

Table of contents

About this book

Everyone ought to know about the Universe, but very few people do. Many of the leading or upcoming citizens of this planet are more knowledgeable on matters relating to stock markets, politics, world cups, even movie stars, than they are about the Galaxy in which they live. The problem is that a lot of the "territory" is unfamiliar, and concepts like the expansion of the Universe somewhat difficult to grasp.

Hence a challenge. Hence this book. It aims to convey an understanding of the Universe, on its largest scale, as clearly as I can manage. Technical terms and scientific wording are avoided. Plain English and common sense prevail. No prior knowledge is assumed.

Acknowledgements

Once again, it has been a pleasure to work with Clive Horwood and John Mason of Praxis Publishing. I am grateful to Stuart Clark and Fiona Gold for feedback on an early version of the text. I would like to thank Tom Matthews, who did much to improve the "readability" of the text, and identified occasional portions that originally proved too difficult to understand or follow. I have also consulted with various specialists, particularly George Ellis and George Branch. I am grateful to my wife for numerous editorial corrections.

Most of the black and white diagrams that accompany the text were created by Margie Walter of the Planetarium, South African Museum, with whom I have long interacted on all matters graphical, and whose advice has always been welcome.

The Universe around us, in three-dimensions, which appears as a colour section, has formed a complementary project to the writing of this book. The three-dimensional starfields have only been possible with the recent database from the remarkable Hipparcos Satellite; I should like to thank Chris Koen (South African Astronomical Observatory) for extracting the star data used here. My close colleague Patrick Woudt (University of Cape Town), helped me in obtaining the galaxy database from the NASA/IPAC Extragalactic Database (which is operated by the Jet Propulsion Laboratory, California Institute of Technology). Heinz Andernach (University of Guanajuato, Mexico) provided the database for the clusters of galaxies. I am, however, especially grateful to Dennis Burford of the Department of Computer Science at the University of Cape Town, who converted my plot files into the colourful visual representations. The final presentation has made extensive use of Corel

Photo-Paint. Herschel Mair of the Planetarium (South African Museum) also assisted. Joanne Young of Audio Visual Imagineering (Orlando Florida) long ago introduced me to ChromoDepth(tm) spectacles, and encouraged me to develop visual materials – the material in this section also forms the basis for a three-dimensional planetarium presentation.

Toward the end of 1999, Prof. Michael Bessell (Australian National University) visited Cape Town, and presented a new set of colour astrophotographs in a talk he entitled "Beauty and astrophysics". I was so taken with this new look at the Universe that I am delighted to include a sample in this book. Others involved in the production of these photographs include Michelle Buxton, Bob Watson, John Shobbrook, Paul Price, Ken Hargreaves, Ralph Sutherland, Hwankyung Sung and students of the Universities of Sydney, New South Wales and Wollongong.

I have also extracted some visual material from the World Wide Web, particularly the Hubble Deep Space images from the Space Telescope Science Institute, and the telescope photographs used in Chapter 5.

Introduction

This book concerns the Universe on the largest scale – its majesty and mysteries. Yet it assumes no previous knowledge in the subject. If you read it, I hope you will gain some greater understanding and appreciation of the magnificent creation we inhabit – much the same as many have already gained appreciation for the wonderful Earth and the millions of fascinating creatures it carries.

Perhaps some warning is necessary: at first glance, our Universe can seem quite intimidating. Our Sun – king of the Solar System – is only one of a *million million* such suns in the Galaxy. It seems so dominant and important to us, yet as stars go it is nothing special, being but one in a vast swarm of stars that forms our Galaxy.

But it does not stop there. Just as our Sun is such a small part of the Galaxy, so our Galaxy is a small part of the observable Universe. There are at least a *million million* galaxies! Our Galaxy with its *million million* stars is just one of them; it is certainly no more significant to the Universe, than is our Sun to it.

Against these overwhelming numbers, our everyday world seems as tiny as an atom. Everything we hold important, everything near and dear, is no more significant than a single grain of sand, when compared to the Sahara.

The *numbers* may seem bad enough, but the *scale* of the Universe is as shattering – bigger than any civilisation of old could have ever imagined. The diameter of the Solar System is ten million times the diameter of the Earth! The diameter of the Galaxy is a hundred million times the diameter of the Solar System! And the diameter of the entire observable Universe is toward a million times the diameter of the Galaxy!

The numbers and the scale are beyond easy comprehension. So big, they boggle the mind. Just to visualise them, we shall have to search for analogies. The sizes and statistics are almost terrifyingly large, when compared to the comfortable world wherein we dwell.

Yet the contemplation of the Universe on such scales is one of the greatest intellectual exercises. How can we feel so important when we occupy so low a rung on the cosmological ladder? Terrifyingly large as the Universe is, it is nevertheless home to humanity, and we could not exist here if the Universe were not "just right" for us. Consequently there is a relationship between our existence and our Universe. This is known as the *Anthropic Principle*. It is one of the matters we will take up for discussion in the chapters ahead.

The observable Universe – our Universe, or perhaps our bit of the Universe – has a limit to its extent. Remarkably, we find ourselves contained within what seems to be a hollow shell. I have called it the *Cosmic Egg*. It sums up the entire Universe.

Quite why it should be like this – well, that is what this book is all about. I shall leave the explanations to the main body of text – as we reach out and feel the texture of the fabric of the cosmos.

Chapter One

At Home in the Universe

Where do you live? It is a question asked dozens of times, and the answer varies from "just down the street" to one's home suburb, home town, country or even continent. I remember being asked the question many years ago by an American taxi driver. Seeking to keep things simple, I said I lived in Africa. "Is that anywhere near Vietnam?" he responded.

I suppose, in retrospect, I might have answered "planet Earth" – not that I can claim to have met anybody from anywhere else, but it's a matter of scale. Home is where you live. It could be the house where you and your family live, or the town, or it might even be the Galaxy – the city of stars wherein we dwell. Even the entire Universe could, in a sense, be considered as home. In fact, if there is a purpose to this book, it is to make you feel at home in the Universe.

This opening chapter's concern is about *where* you live, and with it, I would hope to get you the reader to break mentally from the confines of the Earth's surface. Yes, I know it's a very nice planet (if we do not mess it up any further), but there's much more to it. Home is not just about where we live. It is also about when we live.

We never had a choice in deciding where in the Universe we should live. Neither did we have a choice as to when we should live. For better or worse, we find ourselves at a particular *place* at a particular *time*. Much argument in this book will centre on that particular place and time – this Universe wherein we live, and the time system whereby it evolves – a cosmic fabric of space and time that is home to our bodies and to our consciousness.

Nevertheless, the "world" to which we have become accustomed – home, office, TV, traffic jams, hospitals, stock markets, country vacations...everything of everyday lives – is but a very small part of the Universe. Like the thin film of bacteria that clings to your left back tooth, we live within a film that clings to the surface of a planet.

We have the audacity to call that planet "Earth" – even though less than a third of the surface is dry land, and "Ocean" would have been a far more appropriate name. We also act as though we own the planet, when we can barely scratch its surface. The interior of the Earth is more remote than outer space – the place we are least likely to ever colonise or even visit. It could hardly be more inhospitable. A mere one hundred kilometres beneath our feet, the temperature has risen to a thousand degrees Celsius. By the centre, it is hotter than the surface of the Sun. We go about our daily lives, blissfully unaware of the furnace beneath our feet.

How big is the Universe outside the Earth? History is full of underestimates – to make the perspective less daunting, and the Earth more important. It is curious that the choice for the majority of primitive nations was to roof the apparently flat Earth with a sky dome, within which lay everything seen in the sky – and outside of which was less of a worry. Only the most enlightened of ancient civilisations – such as the Greeks – even considered that the Earth might be a sphere and that its size might be far smaller than the Universe in which it was contained. In Western European civilisation, in the sixteenth and seventeenth centuries, facing the truth that the Earth was not the centre of everything, proved a traumatic transition that entangled the church with science and politics.

As the knowledge of mankind has stretched further into space, so the Earth's role has seemed to diminish toward insignificance – we have long realised that on a physical scale, the size of the Earth makes it almost infinitely small against the scale of the Universe that surrounds it. The most widely used picture in books and journals is a photograph of a "Full" Earth, taken by the Apollo 17 astronauts, en route home. That picture has brought an awareness of our existence on a finite white and blue globe, afloat in the ocean of space.

Those Apollo missions expanded our horizons. From the Earth, the Moon may seem to shine like a silver globe, but up close it is a world of dull dark grey dust. Human shoeprints were left in the powdery but sterile soil – the first time in the Moon's billions of years of history. Lacking the erosive forces of an atmosphere, those shoeprints are destined to remain unmolested for still further millions of years. The Apollo missions brought about a consciousness that there was a Universe beyond the confines of our Earth.

Our view of the rest of the Universe is screened by the scattered sunlight of the daytime sky, but is clearly visible on a cloudless night. The most obvious of observations is that it is generally dark. Trivial though such a fact may seem, we will see much later that it has profound implications on the extent of the cosmos.

1.1 Our Solar System – the Sun and its family of planets – will be familiar to most readers

Specks of light, the stars, punctuate the blackness. Each is an oasis of luminosity. It is at such an oasis that we ourselves live.

Our own star – the Sun – lights up our little corner of the Universe – and is necessary for our existence. We need a place that is neither too hot nor too cold, so a planet that orbits neither too close to the Sun, nor too far away, is required. The Earth is ideal for our existence.

Our Solar System is the one bit of astronomical "turf" familiar to the public (see Figure 1.1). Many people can recite the names of the planets. In essence, the Sun has four small planets – Mercury, Venus, Earth and Mars – that orbit relatively close in. The Asteroid Belt lies just outside these four; its contents sometimes spill inwards, and on occasions collide with the inner planets. Then, at much greater distances from the Sun are four large planets – Jupiter, Saturn, Uranus and Neptune. The Kuiper belt and the Oort cloud lie outside these four; their contents also sometimes spill inwards, and on occasions collide with the planets.

The word *asteroid* means "resembling a star", while "planetoid" might have been more appropriate. The contents of the Asteroid Belt range from minor planets a thousand kilometres across, down to bits of grit. Fortunately for us, it is the smaller pieces that usually spill inward.

What about Pluto? Its status is causing much debate in the astronomical world. When first discovered, it was believed to be bigger than Earth, and so it was accorded the status of *ninth planet*. Today, we know it to be far smaller than our Moon, and simply the largest of a number of similar objects in the Kuiper belt. Had it been discovered nowadays, it is doubtful that it would carry quite the same importance. So it remains as either the smallest major planet, or the largest minor planet, in the Solar System.

Not that the planets are the only planet-sized objects of the Solar System! Jupiter, Saturn, Uranus and Neptune each possess a system of moons – each a miniature Solar System in itself. There are a multitude of large and small moons that orbit the planets, and some are even larger than planet Mercury. Our Moon, for one, is exceptional. No other inner planet has a moon so large. The Earth and Moon are often described as twin planets.

The "planetary" bodies of the Solar System – both the eight major planets and the largest of moons exhibit a remarkable diversity in their surfaces and appearances. The oversize outer planets do not even have proper surfaces. Being composed predominantly of hydrogen, they are often labelled *gas giants*, but the name is misleading as much of their material is more in liquid form. Those that do possess solid surfaces exhibit many of the features found on Earth, but the continents of Earth suffer from drastic water erosion. Without such erosion, the most ubiquitous feature of the planetary surfaces in the Solar System is impact cratering – from the sweeping action that the larger bodies impose on the smaller fry of the Solar System.

The Sun is, of course, considerably larger than any planet – over a hundred times the diameter of Earth for instance – and its greater mass has sparked off nuclear fires in its very dense and ferociously hot core. Its nuclear power provides heat and light to the Solar System. That which bathes the Earth is but a trickle from a phenomenal reservoir. The visible "surface" of the Sun, at 6000 degrees Celsius, is but luke warm in comparison to its interior. The pressures and temperatures are such that the Sun is gaseous throughout.

If our Sun is a star, then the multitude of stars in the night sky suggests that our oasis of light, our Solar System, is hardly unique, but is repeated many times over. There must be solar system after solar system out there. Yet the stars in the night sky are barely the beginning. The naked eye may see a few thousand stars, but a modest telescope sees millions, and large professional

1.2 Our Solar System is but one of many. Note, however, that the true separations between the systems are vastly greater than those suggested in this figure.

telescopes can register billions. Since astronomy now understands that a lot of stars are intrinsically faint, and therefore cannot be seen from great distances, even with powerful telescopes, the total number of stars must be – well – astronomical!

It is astonishing how few people on Earth realise that the Sun is a star, and that the stars are suns! Yet this is clearly the first step toward a better understanding of the Universe. The problem is they look different, and therefore almost every language uses different words to describe them. But ask a child, and he or she will invariably tell you that the Sun is round, and that stars have points sticking out. The idea that stars have "points" is at best an illusion caused by the optics of our eye or a camera. Unfortunately most cultures have entrenched that illusion – five or six-pointed "stars" abound in illustrated literature. One of the first shapes a child is taught to recognise is a "star". No wonder that later on, the same child does not relate the stars of the night to the Sun of the day. And many children still attain adulthood without this misunderstanding being cleared up. Consequently, I would guess that the majority of inhabitants of Earth do not realise that the stars in the night sky are no different in nature to the Sun – basic knowledge we ought to possess.

It is, of course, the distances involved that diminish the stars to pinpoints of light. Looking at even the nearest star in the night sky is like looking at a small coin from a distance of 5000 km. You just cannot see any size to it at all. Stars have such small angular extents that not only can the eye not resolve them, but also the same is still true even when magnified by conventional telescopes. I remember that as a child I was disappointed when looking through a telescope to see the stars as merely brighter pinpoints!

Were it possible to fly with the freedom of a god and approach any of the stars in the night sky, then as one drew closer, each pinpoint would grow to a sphere – an incandescent orb like our Sun, and one would be bathed in light – another oasis of luminosity.

Our Sun is the star of the daytime; the stars are the suns of the night.

Yet stars are often very different to one another, in size and, especially, in luminosity. If we take our Sun for comparison, then there are some stars with several hundred times its diameter (enormous swollen *Red Giant* stars), and there are stars a hundred times smaller than the Sun (incredibly compact *White Dwarf* stars). In light output, the contrast is even greater – a few rare stars have almost a million times the luminosity of the Sun. Yet others may be mere stellar glow-worms, a ten thousandth the luminosity of our Sun, or even less.

Due to this great range in luminosity, there is no way one can tell a star's distance, simply by judging its brightness. Of the brightest stars in the night sky, around half are bright because they are nearby, neighbours to our Solar

System. The others are not nearby, but incredibly luminous – so even from a great distance they still appear bright enough to rival our neighbours in the night sky. Similarly, the faintest stars visible in the night sky are either stars that are very distant or nearby stars that have very low luminosity.

It is then easy to imagine that the Universe in which we dwell is simply full of stars – as though the starry heavens continued almost *ad infinitum*. That indeed was the generally accepted picture at the beginning of the twentieth century. But already then, some farsighted individuals were advocating the existence of "island universes" – as though our "Universe" of stars was repeated many times over.

Indeed they were right – the island universes are now known as *galaxies*, and like islands in an ocean, featureless oceans of apparently empty space separate them. But separate universes they are not. We talk today of our Universe being populated by galaxies, and though nobody yet possesses a spacecraft capable of travelling to our nearest neighbouring galaxies (and probably they never will), we can look with our telescopes and see the other galaxies – much like our Galaxy of stars, but repeated many times over.

The term *island universe* is more like something the advertising agencies would use. It is so appropriate. It conveys the idea of an unbelievable number of stars in a gigantic system – and that is a good quick description of *our Galaxy*.

Picture an immense disk of luminosity, so huge that it would take a ray of light a hundred thousand years to cross from one side to the other. Close inspection shows that the luminosity is derived from a multitude of points of light – the stars! It is a city of stars – but with a population vastly exceeding the population of a terrestrial city. There are estimated to be a million million stars (an American *trillion)* in our Galaxy – a truly phenomenal number. We should gasp at the thought – there could be a million million solar systems such as our own in this island universe. Suddenly, our own seems *very* insignificant.

The analogy to a terrestrial city is a good one. Most terrestrial cities consist of a busy city centre, probably with tall buildings, surrounded by a disk of suburbia three-dimensional in the centre, and mainly two-dimensional in the surroundings. The same is true for our Galaxy. It has a three dimensional "central bulge", packed with billions of stars, that forms the central hub to a gigantic disk – somewhat reminiscent of the planet Saturn, but with a more extensive ring system. Figure 1.3 gives an impression of the geometry.

The disk of the Galaxy carries a spiral structure, with distinct spiral "arms" winding out from the central hub. The whole affair looks remarkable similar to a "catherine wheel" of a fireworks display. The spiral pattern suggests rotation, and indeed this is so – the disk rotates slowly, the inner part revolving

1.3 Our Galaxy

faster than the outside. The stars in the disk orbit about the central bulge, much as the planets of our Solar System orbit around our Sun. But since the scale is so enormously different, so is the period of revolution – in the region of 100 million years!

Our Sun does not hold any special status in our Galaxy. Though brighter than average, there are many stars that could outshine it. Given that some are nearly a million times brighter, the Sun would hardly be noticeable if the Galaxy were viewed from afar; indeed it would be totally lost amongst the multitude. It is those stars, much brighter than the Sun, that define the spiral arms. The spiral arms stand out, not because they contain more stars than their surroundings, but because they contain brighter stars than their surroundings.

Our Sun does not hold any special place in the Galaxy. We do not even dwell in the busy central bulge - but are relegated to the outer suburbs, far out from the city centre, where it takes around 200 million years to orbit the Galaxy. More precisely, the Sun lies toward the inner edge of one of the spiral arms - the local arm or *Orion Arm*. There is some doubt as to whether this local arm is a proper spiral arm, or only a *spur* from a spiral arm. Whatever, it is not a permanent address - the nature of the spiral structure is such that the Sun will in time pass through the arm, and eventually drift through other arms, as indeed it must have already done in the past.

Our Sun therefore lies within a flattened disk of stars, but displaced far from its centre. Because the stars are concentrated toward the plane of that disk, and because we ourselves see the Galaxy from within, the disk appears as an encircling band of luminosity. We know it as the Milky Way. The Milky Way that we see in a dark night sky (without city lights or moonlight) is simply the local portion of our Galaxy stretched around us. Only in one part of the sky, where the Milky Way seems broader, do we see beyond our local portion of the Galaxy, to part of the central bulge itself.

Our view is confined to a local part of the Galaxy because stars (and planets) are not the only constituents of our Galaxy. Seen in silhouette against parts of the Milky Way are dark clouds of interstellar matter - apparently thick enough that you cannot see through them. Although 99 percent of this interstellar matter is gas (generally cold and transparent) the smaller content of particles (simply dust) is sufficient to make the clouds completely opaque to optical light. There are enough clouds for our view of the Galaxy to be limited to relatively short distances. It is like being in a forest where the density of trees allows you to see only a limited distance, so you see but a small part of the forest and not its entirety.

The gas and dust clouds tend to congregate in the spiral arms, particularly in the regions of compression along the inner edges. The most conspicuous dark lane, seen against the Milky Way, lies along the inner edge of our *Orion Arm*. It is known as the *Great Rift*, because it appears to divide the Milky Way in two. Though these dust clouds interfere with our view, we can nevertheless look out of our spiral arm. When we look inwards in the Galaxy, we see the neighbouring spiral arm - the *Sagittarius Arm* – whose dust clouds block our view further, in the plane of the Galaxy. Similarly, looking outwards in the Galaxy, our view ends with the dust clouds of the next spiral arm out - the *Perseus Arm*. In short, all we see are fragments of three spiral arms, and a bit of the central bulge.

Aside from stars and dust clouds, there are also glowing emission clouds to be seen. This is where the interstellar gas - normally cold and transparent - has been excited by ultraviolet light from particularly hot stars. Usually, the

1.4 Our Galaxy is but one of many. However, the separations between galaxies are somewhat greater than those suggested here.

clouds give off the reddish light of the hydrogen emission spectrum, since hydrogen is by far the most abundant element in interstellar space (and the Universe in general). The mixture of star fields interrupted by dust lanes and glowing clouds makes for impressive and beautiful scenes in our local portion of the Galaxy.

If our Galaxy were the entire Universe, then the Universe would indeed be a wondrous thing; but just as our Solar System is but a tiny part of the Galaxy, so the Galaxy, despite its immensity, is but a tiny part of the observable Universe. If one looks out, above or below the plane of the Milky Way, one sees out into extragalactic space. Out there, we see other galaxies, many similar to our own. They are as numerous as the flowers that bloom in the Spring, and are as common in the Universe, as were the stars within our Galaxy. The magnitude and scale of the observable Universe is indeed awesome!

Many of the other galaxies look similar to our Galaxy – though galaxies, like humans, are never exactly alike. But the blueprint that produces spiral galaxies has been distributed on a universal basis. Countless spirals, twirling this way and that, adorn the cosmos like stupendous Yuletide decorations. Not that all galaxies are spirals. Some galaxies, particularly those that inhabit denser clusters, have the "central bulge", but lack a disk or any spiral structure. These are known as *elliptical galaxies.* There are also smaller so-called *irregular galaxies.* (Again these can be seen in the photographic sections of this book.)

As galaxies go, ours is a moderately large spiral galaxy, but there is an even bigger one as a neighbour. Both the *Great Galaxy in Andromeda* and our Galaxy are accompanied by a small swarm of satellite galaxies. For our Galaxy, the most significant of these are the two Magellanic Clouds, seen as patches of luminosity in the southern sky, and the Sagittarius dwarf.

At first sight, the galaxies populating extragalactic space, with plenty of apparently empty space between them, look much like the stars populating our Galaxy, which have plenty of space between them. It is just that the whole scale is different. The spacing between galaxies is almost a million times greater than the spacing between the stars in our Galaxy. However, relative to the distances that separate them, the sizes of galaxies are larger. In photographs, you can identify them as galaxies, whereas stars are seen as pinpoints of light. Galaxies have to be very distant before they appear as little more than blurs of light against the blackness of the sky.

Nevertheless, it is the blackness of the background sky that dominates – not the luminosity of the galaxies. The Universe, in general, is a very dark place. If one were simply thrust at random into the Universe, somewhere outside the comfort of our home Galaxy, it would look for a moment as if we had been put into absolute darkness. Once dark adapted, our human eyes have just enough sensitivity that we would just be able to make out a few of the nearest galaxies – but that would be all. The Universe is dark, much like the background night sky seen from Earth. The only exceptions would be if we were to equip our eyes with a telescope, or image intensifier – some device that gathers much more light than the small pupil of the eye then, of course, we would see much more (even enough to write this book all about it). Alternatively, if our eyes were sensitive to electromagnetic radiation outside the wavelengths of visible light, then we would see luminosity all over the sky in *microwaves* (but the explanation of this forms the topic of the next chapter).

There is another big difference between the Universe of stars within our Galaxy and the Universe of galaxies; it is in the way they are distributed in space. Though stars form occasional clusters, they are scattered almost randomly throughout the volume of the Galaxy. Not so with galaxies. An astonishing revelation of the late twentieth century is the way in which galaxies are distributed in space. Instead of being distributed randomly throughout space, galaxies congregate into large-scale structures – a labyrinth of interconnected filaments and "walls" of galaxies. The texture of this labyrinth is not dissimilar to that of a sponge, so that numerous voids are embedded within it, apparently devoid of galaxies. The "bath-sponge" texture is on so large a scale as to be almost frightening – something that might have seemed more at home in the cyberspace of a computer than reality on the largest scale possible.

The cosmic labyrinth obviously conveys something fundamental about the texture of the Universe – a message that scientists have been keen to discover ever since it was revealed. It is the fundamental framework of the Universe, from which galaxies, stars, planets, and even we ourselves have consequently evolved. (Chapter 4 will be devoted to its consideration.)

Our Galaxy is a part of the cosmic labyrinth. Our Galaxy and the Great Galaxy in Andromeda, with lesser companion galaxies, form a condensation known as the *Local Group*, which is in turn part of a protuberance of the local *supercluster,* centred on the *Virgo Cluster* of galaxies. The local supercluster is in turn an appendage of a great flattened structure in the general direction of the constellation of Centaurus.

Currently, our mapping of large-scale structures extends only limited distances outward. Much more detail as to how this was achieved will be described later. Presumably the cosmic labyrinth extends to ever greater scales.

This, then, is our cosmic home. Our Earth is but one planet circling our Sun. Our Sun is but one star of billions circling our Galaxy. Our Galaxy is one of thousands assembled into the great array of local large-scale structures. All this is home. Our own Earth may for the moment seem quite insignificant against the immensity of the Universe, but this will prove far from the case (as we will discuss in later chapters).

The diagrams incorporated in this chapter have necessarily been schematic and conceptual in nature. However, the first colour section included in this book presents photographs of the Galaxy about us. They capture something of the beauty of the neighbourhood within our own city of stars, before moving on to show pictures of other galaxies. In essence, the pictures are a portrait of our home in the Universe.

Although we can look out at the stars and galaxies, we can never expcct to travel around our Universe. The scale is simply too enormous. Even other stars are too distant for us to reach at present. Other galaxies are even more so. Yet, we can sit and look ever deeper into the cosmos. One is tempted to think, as was thought long ago of the universe of stars, that the same would continue over and over again on an endless, or almost endless, scale.

That is not the case. The entire observable Universe is only about twenty times larger than the biggest of the large-scale structures. Perhaps more dramatically, we see a boundary to the observable Universe – a limit to all that can be seen, a limit to all that will ever be seen. In spite of the tradition, that given ever more powerful telescopes we would be always able to see deeper into space, this is no longer the case. In the twentieth century, we saw the boundary of the observable Universe – for the first time in the history of the human race. This is not to say that we have seen everything in the observable Universe; there is an enormity of constituents still to be examined, for which ever more powerful telescopes will be required.

But the idea that there is indeed a *boundary* to the observable Universe is something that takes some time to accept, some time to digest, and some time to explain. It is the topic of the next chapter.

Chapter Two

Living Inside the Cosmic Egg

Imagine waking up one morning and finding yourself floating at the centre, inside a *hollow chocolate Easter egg!* Its shell surrounds you. You are trapped. You cannot see out, and you cannot get out. Worse still, particularly if you are a chocolate addict, the walls are too far away to get a bite. Like Tantalus, you stretch out, but the walls of the egg are perpetually out of reach.

It may sound ridiculous; nevertheless it is not far from the truth. It is the view we get of the surrounding Universe. *We find ourselves contained within a very large spherical shell.* We cannot get out, and we cannot see out. A *Cosmic Egg* wherein we live for our eternity. An egg that, like its chocolate counterpart, never hatches. We have no way of ever knowing what is outside.

Remember that the Universe is so vast that we cannot travel about within it. But we can nevertheless see it, and the Cosmic Egg is what we see. In reality, there is no physical structure to it – but it appears as real as can be. This chapter is devoted to explaining how this comes about.

When we look at the surrounding Universe, we see *light* (or, with sophisticated modern technology, other types of electromagnetic radiation – infrared and so on). The crucial point is that light (or any other form of electromagnetic radiation) only travels at a *certain limited speed.*

Of course, compared to everyday life, travelling at the speed of light is very very rapid, and so it is, 300,000 kilometres per second or just over a billion kilometres per hour. (The "billion" used in this book is the more familiar American billion – a thousand million – it is also a convenient and concise unit for matters cosmological.) But, astronomically, light is not particularly fast. For a light ray to travel from the Moon to the Earth, it takes a little over one second. Because of this, when we look at the Moon, we see it – not as it is this instant – but as it was over a second ago. It takes light over eight minutes to travel from the Sun to the Earth. Consequently we see the Sun as it was over eight minutes ago. If the Sun could suddenly be switched off (highly unlikely to say the least), we could go on enjoying its light and warmth for over eight minutes, until the news caught up with us!

By the same argument, we see planet Jupiter as it was about 35 minutes ago. When in 1989, we sent the *Voyager 2* spacecraft past planet Neptune, we had to wait patiently for four and a half hours to receive the pictures it transmitted. That spacecraft (at the time of writing this book) is now about ten *light hours* out. Its radio transmitter is still working, but two way communication now takes about twenty hours!

And there is nothing we can do about it. We might prefer to have instantaneous communication with our spacecraft, and we might prefer to see the planets as they are at this moment, and not as they were hours ago. But we cannot change that most fundamental law of the Universe – *nothing travels faster than light.*

It may take light over four hours to travel from the outer planets, but from the nearest neighbouring Solar System – the Alpha Centauri system – it is *over four years!* Such is the contrast between the distances to the planets in our Solar System and the distance to other Solar Systems. The Voyager 2 spacecraft will eventually make it to some other Solar Systems, but only over millions of years of travel time. For now, we cannot easily conceive how we humans will travel to the stars – though ideas of suspended animation and travelling colonies abound in the realm of science fiction. We can only look at the stars and see them as they used to be – even that nearest neighbouring system is seen as it was over four years ago.

Step outside one clear night and take a look into history. If light from our nearest stellar neighbour takes some four years to reach us, then the light from most of the bright stars you see in the night sky takes somewhat longer – often tens of years, and sometimes even hundreds. None of the stars that you see tonight are as they might be at the moment – only as they were back in history.

Alpha Centauri is seen from the Southern Hemisphere of Earth, and is one of the "Pointer" stars alongside the Southern Cross (see Figure 2.1). If you are able to see it tonight, then what you see is the star as it was four years ago. But if you wanted to see it as it *is* tonight, you would have to wait over four years to do so, and more than likely, it would look exactly the same! In popular notation, astronomers often refer to Alpha Centauri as being over four *light years* away – a representation of both its distance and the "look-back time" to that star.

The other "Pointer" star, Beta Centauri, looks as though it was a neigbour to Alpha Centauri – but it is really a line of sight effect. In truth, it is a much

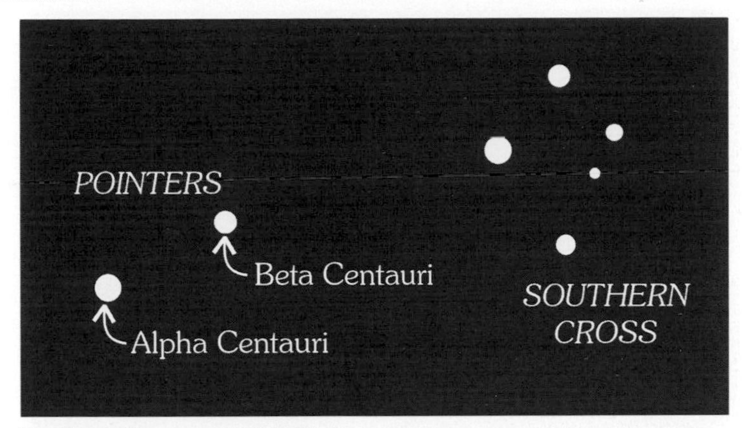

2.1 The Southern Cross and Pointers

brighter, much more distant star – 525 light years away. So the Beta Centauri you see tonight is Beta Centauri in the late 14th century – as it was just before Columbus set out for America. Similarly, the four bright stars that form the Southern Cross are a few hundred light years away, and you see them as they were a few hundred years ago.

In a different part of the night sky, *Orion* is probably the easiest of constellations to identify (see Figure 2.2). It has a distinctive line of three bright stars – *Orion's Belt* – like nowhere else in the sky. It happens that these three stars are incredibly luminous and consequently about the most distant individual stars the eye can see for instance, Alnilam, the central star of the three is seen in 650 A.D., at a time when Christianity was first introduced to Britain.

The naked eye can still see further back into history. The diffuse light of the Milky Way comes from countless faint stars, too distant to be seen individually. Like the glow of a distant city at night – too far away for individual streetlights to be visible – the stars combine their light into a luminous cloud. They too are part of a city; our Galaxy, the Milky Way, is simply our city of stars, seen from within. Only a few thousand relatively nearby stars within our Galaxy are bright enough for our eye to see individually, the vast majority is too faint or too distant.

On average, the glow of the Milky Way comes from stars at distances of a few thousand light years (still relatively nearby in the Galaxy) so when you see that glow you look back to the time of the building of the Egyptian pyramids. However, in the constellations of Scorpius and Sagittarius (Figure 2.3), we see a more distant portion of the Milky Way – the central bulge of our Galaxy. Since this is some 30,000 light years out, we see back to almost 30,000 BC – far before recorded history, when the human race had yet to migrate over the entire planet. Few people realise how far into history the naked eye can penetrate.

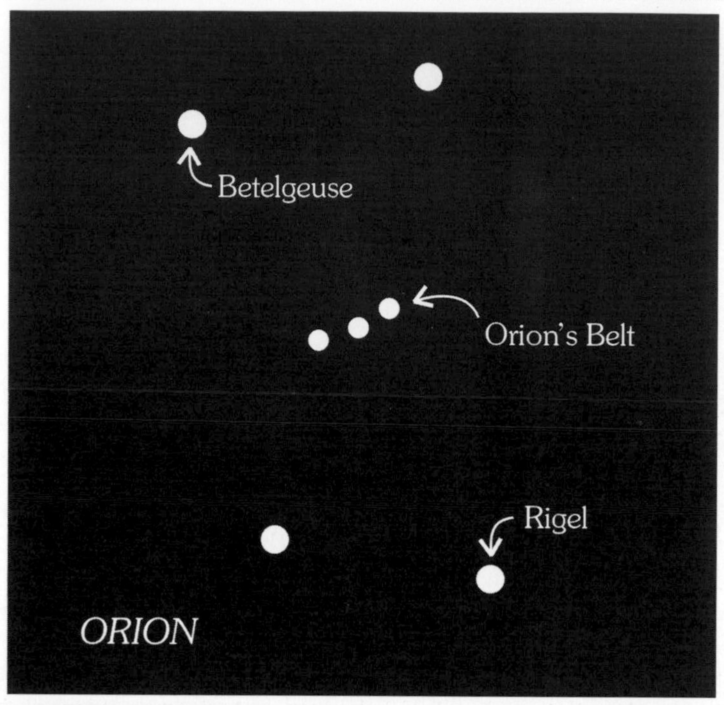

2.2 The constellation of Orion

2.3 The constellations of Scorpius, Sagittarius and the Milky Way

To help you assimilate this perspective of looking deeper into space, the second colour section in this book features *The Universe around us, in three dimensions.* It consists of a sequence of three-dimensional diagrams that progress deeper and deeper into space. They allow you to reconcile the objects in the night sky – such as the planets, the stars and the Milky Way – to their distances, expressed in light years. For instance, one of the views spreads the distance perception from nought to a thousand light years – so that the starfield normally visible to the naked eye is seen in its true perspective.

The central bulge of our Galaxy is still not the limit to which we can see on a good dark night. The unaided eye can discern three neighbouring galaxies to our own. Two are the Magellanic Clouds, only visible from the Southern Hemisphere of Earth, and seen as diffuse luminous clouds – like the Milky Way, but separated from it. The Larger Magellanic Cloud, the nearer of the two, is some 160,000 lights years away, the Small Cloud about 180,000 light years, so we see back the same in years into history.

The third galaxy is the record holder, the most distant object normally visible to the naked eye – the Great Galaxy in Andromeda in the northern sky. Admittedly, the eye sees but the central bulge as a faint elongated blob – but the distance is two million light years away! And so we see that galaxy as it was two million years ago. We are looking back to a time when *Homo Habilis* – a forerunner of modern man – was emerging from the evolving species on this planet.

We have no way of seeing the present astronomically. In one sense, this is not a problem since stars usually endure for millions of years, and even if we could see the Great Galaxy in Andromeda as it is today, then it is unlikely to have changed very much. But the idea that we are forced to look back into the past – and cannot see the present – is obviously an important understanding of the *view of the Universe* we have today.

We can't help but wonder if the Universe today is in any way different to the Universe we see tonight. Are all those stars still shining? Have not any gone out? There is at least one that has blown up. It is one of the stars in the Large Magellanic Cloud – far too faint to see with the naked eye, but bright enough to have been catalogued and classified. Except it was not there. Way back around 160,000 BC it blew up in a catastrophic *supernova* explosion. With the possible exception of the very core of the star collapsing, most of the star's content was torn apart and scattered far afield. But the news only reached us in 1987! Prior to that we have been looking at the image of a star that no longer existed. A star long gone, but its light lived on.

Of course, these things happen more readily in science-fiction movies – long long ago in galaxies far far away. But, even if the whole Great Galaxy in Andromeda had been destroyed by some monstrous evil empire a million years ago, the news would only reach us in over two million years time!

To see the Universe on a cosmic scale, is to dwell in history.

The naked eye might be able to see back two million years, but equip it with a modern telescope, and it will do astonishingly better. Telescopes have enabled us to see history as never before! The power of a telescope lies in its ability to capture much more light than can the tiny pupil of the eye. More light means being able to see much fainter objects. Within our Galaxy, telescopes see individual stars much more distant that those in Orion's Belt. Outside the Galaxy, they see other galaxies (as again can be seen in the colour section *The Universe around us*), and the distances are awesome.

They sight galaxies that are many millions – even billions of light years distant. And in doing so, they look back millions, even billions of years in time! This is history way beyond that of humans – akin to going back to earlier ages of the Earth. We can set a perspective by noting that the age of the Earth is four and a half billion years, and that the age of the Universe is probably just less than 15 billion years. Modern telescopes see back way further than the age of the Earth and through most of the history of the Universe!

So when astronomers use telescopes, they are exploring deeper into space, and also deeper back in time.

It is quite incredible to realise that we can sight galaxies tens and hundreds of millions, and billions of light years out. And the same number of years back into the past. We are able to look back through the history of the Universe. We cannot see our own Galaxy – or any of our "local" surroundings – as they were in the past, but we can see other galaxies. (We shall cover more detail in Chapter 5.)

How far out we can see in space depends on *how far back we can see into the past*. Had the Universe always been in existence, then we might have been able to see "forever". But that is not so. There is ample evidence that the Universe, as we know it now, has not existed forever (as will be discussed in the next chapter). It has a *finite* age. Although in that age of the Universe, a light ray could travel an immense distance, it is nevertheless not an *infinite* distance. If we take the age of the Universe as around 15 billion light years, then the furthest any light ray could travel is simply 15 billion light years. It is an enormous distance. It is the record – no light ray has travelled further because the Universe is no older. *It is the furthest we can hope to see in the Universe.*

Even if the Universe is infinite in extent, we can only see a limited distance, and a limited volume of the Universe – effectively a sphere centred on the Earth. That is the entire *observable* Universe – all the Universe we shall ever know. It is not a matter of technology – bigger better telescopes can never see further.

But there is a complication. For light to travel unimpeded through the Universe, the Universe has to be *transparent*. Though extragalactic space is dotted with galaxies, the dots are very far apart, so extragalactic space is by and large transparent. But – and here is the crunch – this has not always been the case.

NEARBY IN OUR GALAXY

The well known constellation of Orion is recognisable in this wide-angle photograph. The exposure reveals far more than the naked eye could ever see, particularly the luminous gas clouds (in red) – one around the star Lambda Orionis (left), Barnard's loop and the Great Nebula (the brightest). (Unless credited otherwise, all photographs in this section are from Prof. M. Bessell and collaborators – see details in Acknowledgements.)

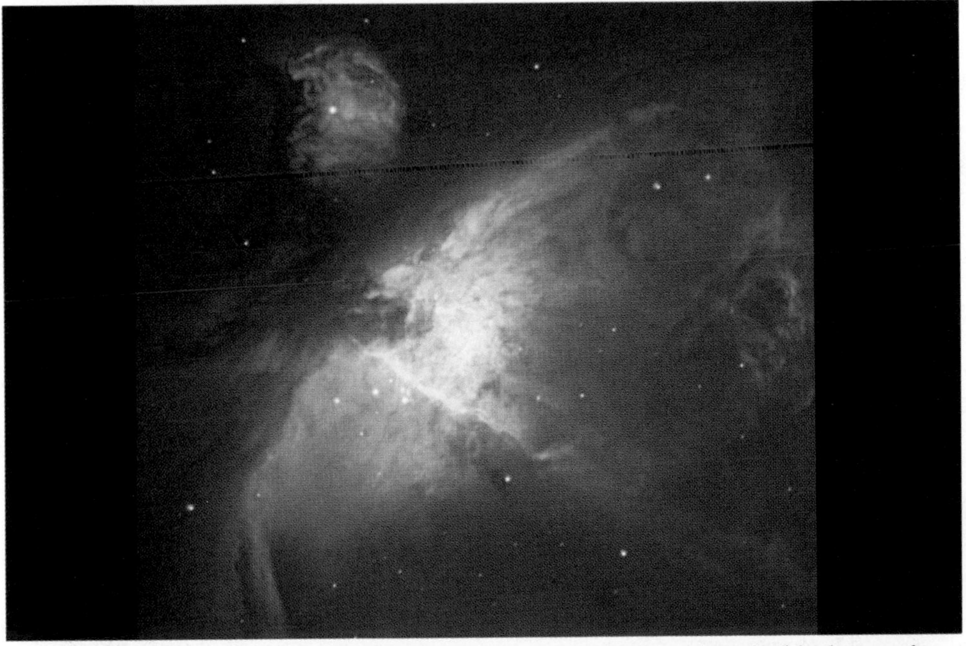

A close-up view of the Great Nebula in Orion, the glowing gas is excited by hot newly formed stars – effectively a "maternity home" in our local spiral arm in the Galaxy.

Another wide-angle photograph shows the Southern Cross and Pointers, but reveals both luminous gas clouds and dark clouds, seen in silhouette against the starry background of the Milky Way.

A close-up of the Eta Carina Nebula (near right hand edge of previous photograph), centred on an extraordinary star seen to erupt some one hundred years ago.

IN OUR GALAXY

Hot young stars have created a cavity in the centre of the Rosette Nebulae, their ultraviolet light exciting the gas beyond.

Another nebula excited by a hot young star. The dust lanes seen against it lend to its being named the Triffid Nebula.

A dying star gently blows its outer layers off into interstellar space, exposing its very hot core (the starlike object in the centre).

Not all stars die peacefully. The Crab Nebula is the remant of a massive star that became unstable and exploded.

OUR NEIGHBOURING GALAXIES

This wide-angle view shows our two neighbouring satellite galaxies, the Large and Small Magellanic Clouds. The prominent red cloud in the Large Cloud is known as the Tarantula Nebulae.

A close up of the Tarantula nebula, with colours adjusted to enhance details in its gaseous content, revealing an almost three-dimensional texture.

OTHER GALAXIES

A magnificent spiral galaxy, NGC 2997, seen almost flat on. The colour balance used for the pictures on these pages shows both stars (diffuse blue light) and glowing clouds (red).

NGC 6744 is another spiral galaxy seen flat on.

OTHER GALAXIES

This galaxy, NGC 1365, shows a straight bar in its central region.

Known as the "Antennae", this peculiar object is the outcome of two galaxies in near collision with one another. Tidal turmoil has thrown out long tails.

THE DISTANT UNIVERSE

By far the deepest photographs ever taken, the Hubble Space Telescope probed the distant cosmos in two opposite directions (Space Telescope Science Institute).

First consider an example. We look up one day and see a blue sky dotted with cumulus clouds. In between the clouds, the air is quite clear – quite transparent. But the clouds are not. You cannot see through them, you cannot see into them. Of course, if you boarded a plane, you could fly through them, and once inside a cloud, you would find it like a fog. You could probably see the wing tips of your aircraft, but not much further. From the outside, the cloud looks like a three-dimensional body floating in air. It seems to have an outside surface, but the surface is not abrupt – you can see a wing's length or so into the cloud. But otherwise the clouds are quite *opaque*, whereas the air in between is *transparent*.

Now let us look at a similar example – the Sun. The Sun is an incandescent ball of gas. But the gas is generally *opaque*. You cannot look into the Sun. It comes about because of the physical conditions; gas that is dense and very hot is generally "ionised" and that makes it opaque. In fact were it possible to fly a plane into the Sun, you would find that you could see much more than a wing's length; you would be able to see about a hundred kilometres. Even so, that is not very far, compared to seven hundred thousand kilometres – the radius of the Sun. So as with the cumulus cloud, the Sun is *opaque*.

But not completely. The Sun looks like a ball – as though it had a surface to it, as did the cumulus cloud. The apparent surface is known as the *photosphere*. But the photosphere is not the boundary of the Sun! There is still gas above the photosphere. The difference is that the gas above the photosphere is transparent, whereas the gas below the photosphere is opaque. Why is this so? It follows from laws of physics – hot denser gas is opaque, slightly cooler less-dense gas is transparent.

The third, and final example, is ... the Universe.

In the next chapter, we shall explore the evolution of the Universe according to the *Hot Big Bang* – the theory overwhelmingly favoured by cosmologists. It holds that, in the beginning, the Universe was very hot and very dense. The early Universe was very different to the Universe today. Galaxies, stars and planets – the stuff in the previous chapter – were yet to form. Instead the entire Universe was full of hot dense gas. And, yes, it was *opaque.*

The entire Universe was then much like the inside of our Sun is today! We couldn't have survived in that early Universe, any more than you or I could live inside the Sun today. But like the inside of our Sun, we would have found the gas to be incandescent and foggy. Again, one would only see a limited distance. The *entire* Universe was opaque, not transparent.

However, it didn't stay that way. As the Universe expanded, it got less dense and cooler. And there came a time when it made the transition from being *opaque* to being *transparent*. It must have been a dramatic event – the clouds suddenly clearing, and the whole Universe becoming clear and transparent.

Our understanding of physics tells us *when* this would have happened. It would be when the Universe had cooled to around three thousand degrees, which would have occurred about 300,000 years after the big bang beginning itself. It would not have been an instantaneous event, but a gradual clearing.

What a sight it must have been to behold the Universe changing from foggy to clear – a true dawning of a magnificent creation. You might think what a pity we were not around then to see it. But that is where you could be wrong. *We can still see it!*

Perhaps one of the most breathtaking facts of cosmology, is that we can look right back to see the Universe at age 300,000 years!

Remember, the further we look out, the further back into history we see – so long as the Universe is transparent. But there comes a point, when we look back to where the Universe is only 300,000 years old! We cannot look beyond, because earlier than that, the Universe was *opaque*. It forms a shell, through which we will never see.

Figure 2.4 shows, in schematic fashion, our view of the Universe, the entire observable Universe. The picture is, of course, three-dimensional. The outer boundary – the shell – is the early opaque Universe. It is as opaque as the shell of the chocolate Easter egg. We cannot see through it, and we will never see though it.

This is what I have chosen to call the *Cosmic Egg*. It is not a physical structure, since it exists in time, not in place. But because we have no choice but to look back in time, it contains us as would the shell of an egg. It is what our eyes would see – were they only a bit more powerful and also sensitive to longer wavelengths of radiation. Light rays cannot penetrate opaque objects, and no matter what equipment we may develop in the future, we will never see through the shell. We will never see out of our Cosmic Egg.

Contained within that Cosmic Egg is the entire observable Universe. All we know and all we shall ever know. The Universe may extend far further in space, but the rest of it lies forever beyond the reach of our telescopes and our curiosity. We shall never see it.

Indeed, we can look back through almost the entire history of the Universe. The only bit we cannot see is that early period, the first 300,000 years, when the Universe was opaque. As long as the Universe has been transparent, all the time ever since, we can look back through it.

The shell of the Cosmic Egg may erroneously suggest that we lie in the "centre" of the Universe. We do indeed lie in the centre of the *observable* Universe – but that is no more than the effect of an horizon. Another observer situated in another galaxy would have another cosmic egg – centred on his or her galaxy instead. The situation is no different to a horizon on the Earth. If a sailor looks out from the crow's nest of a ship at sea, he sees the same distance

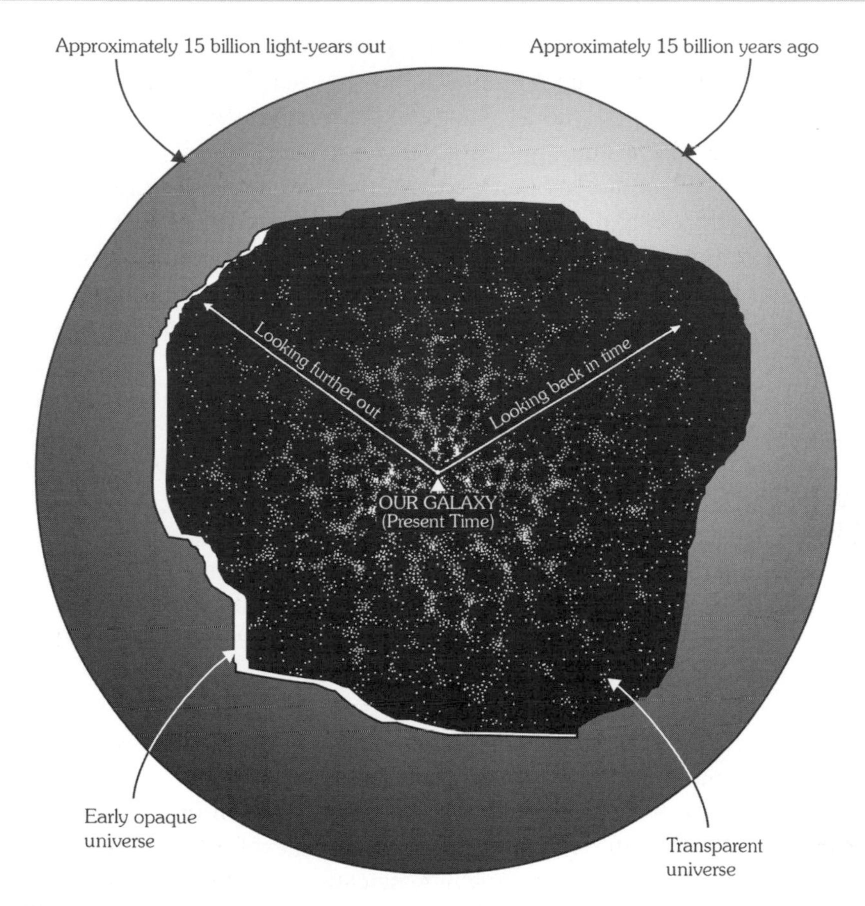

2.4 The observable Universe is contained within the "Cosmic egg".

in each direction. But that does not mean he is in the centre of a circular ocean. As his ship moves, so the horizon moves. On the cosmic scale, the observer cannot move, so the horizon is effectively fixed in space and time. It is simply the limit of visibility and not because we are at any "central spot" in the Universe. We shall see in the next chapter that there is, anyway, no central point.

In fact, an astute reader may realise that the Cosmic Egg is nevertheless growing larger as the Universe gets older. As time passes, the radius of the observable Universe must grow at the speed of light, since the look-back time depends on how far light has travelled in the age of the Universe. Surely, then, we should gradually get to see more? Not really. In the next chapter, we will discuss the expansion of the Universe, a consequence of which is that the shell is moving away from us at so close to the speed of light, that what is inside the shell now remains the same as the Universe expands. What is currently outside the shell shall remain outside the shell.

This then – as represented by Figure 2.4 – is the *view of the Universe* we see today. It is constructed on the simple assumption that light travels in a straight line, and due to the finite age of the transparent Universe, we see only to a limiting distance. Hence the sphere, the Cosmic Egg.

The Cosmic Egg embodies both the past and the present. What we see inside it is the past – the further out from Earth, the further back in time – as emphasized in the foregoing discussion. Yet the radius of the sphere is given by the *present* age of the Universe, not its past.

We look into the past, yet we see it stretched to a present day scale!

What does this mean? It means that when looking further out, and further back in time, what we see becomes ever more distorted by being stretched to a present-day scale. Even the light it emits also gets stretched to a present-day scale. We may be able to see back in time, but as we look a long way out, so the images we see are distorted and their "colour" changed.

Their colour is changed by the process we know as redshifting – something that will be explained in much more detail in the next chapter. For now think of it as simply *stretching*. When the waves of light are stretched, the colour is changed – the light moves toward the red end of the spectrum. Extreme stretching will even move the waves out of the region of visible light – into the *infrared*. Enormous stretching – like making everything a thousand times longer – would shift waves from visible light right through the infrared region, so that they would end up in the *microwave* region.

The Hubble Space Telescope has let us look so deep into space, that the galaxies it sees at extreme distances look somewhat different, when compared to those nearby. While this is partly because we see them at a much earlier epoch, it is also because their light has been stretched, so that the colours are not the same. Astonishingly, we are now able to see galaxies at a time when the Universe was just a fraction of its current age.

But one of the greatest miracles of the twentieth century is that we are able to see deeper still – right to the inside shell of the Cosmic Egg.

A little earlier we likened the shell of the Cosmic Egg, to that of a chocolate Easter egg. This is not a bad analogy as far as visibility by the human eye is concerned. The reader has already, I am sure, realised that you cannot see it when you go outside on a clear night. Like chocolate, it is dark. But, if only your eyes were sensitive to *microwaves*, you would see the night sky not as dark but ablaze with light. The light from long long ago.

Long long ago was when the Universe, at its youthful age of only 300,000 years, made that crucial transition from being *transparent* to being *opaque.* Remember that prior to that the Universe was ablaze with light, like the inside of our Sun. But the light then never survived to this day, because it was constantly being absorbed and re-emitted by the dense gas of the Universe.

However, there came that time, when the gas expanded and became less dense and cooler, and no longer absorbed the light. The Universe became transparent – *and the last light emitted survived!*

It survived such that we can still see it today. But not as visible light. The extreme stretching due to the expansion of the Universe has made it into microwaves. The microwave radiation from the shell of the Cosmic Egg is the most abundant radiation by far in the Universe. The Universe is anything but dark – it is filled with microwave light.

If you could look with microwave eyes, you would see the inside *surface* of the shell of the Cosmic Egg. The shell would appear to have an inside surface, in the same way as the Sun appears to have surface, in the same way as a cumulus cloud appears to have a surface. The surface is the boundary between opaque and transparent material.

It is an amazing achievement that we have, in relatively recent times, seen the inside surface of the Cosmic Egg – the very boundary of the visible Universe. The microwave light that it emits is known as the *Cosmic Microwave Background*. It is the final limit to all we see and will ever see in our Universe.

But, there is still another surprise! This Cosmic Microwave Background carries with it .... a *snapshot of the very early Universe.* Just as the Universe graduated from being opaque to transparent, so it had a "graduation photograph" taken – a record of what it looked like at that crucial point in its history, and an invaluable discovery for the cosmologists of today.

At first sight, the photograph may be unimpressive. After all it comes from a time before galaxies, stars and planets (the stuff of our opening chapter) had formed. The Universe was still much like the inside of our Sun – incandescent gas everywhere. And, like looking at the surface of the Sun, at first sight it seemed featureless – uniform radiation all over.

But not quite. While discovered in the mid 60s, by the mid 70s, a slight brightening of the Cosmic Microwave Background on one side of the sky was apparent – by only 0.1 percent. However, that brightening has since been realised to have nothing to do with the original photograph. Instead it results from our local movement (and will be the basis for discussion in Chapter 6).

Given that the Universe has so many large-scale features today, the absence of corresponding detail in the photograph of the Cosmic Microwave Background was, for many years, a big worry for cosmologists. However, the worry was set aside in 1992 when analysis of data from the American "COBE" (Cosmic Background Explorer) satellite showed a pattern of faint variations at a level generally less than 0.001 percent. Though many researchers were initially skeptical of the claim that the long sought variations in the "photograph" had been found, the care of the researchers and its subsequent verifications have since convinced most cosmologists.

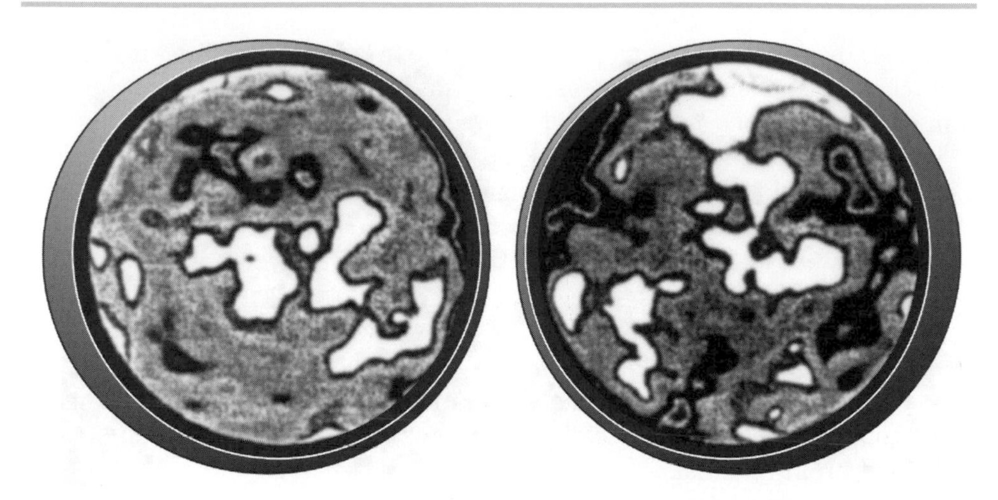

2.5 – The inside shell of the Cosmic Egg carries an image of the early Universe.

Figure 2.5 shows what was found. It is a map of the entire sky in the light of the microwave radiation and a picture of our Universe in its infancy. It is a picture of the inside shell of the Cosmic Egg. The picture is not of very high quality, in so far as the *resolution* of the camera was several degrees in angle – like that of a camera that produces a very blurred picture. It does however show that some patches of the inside shell of the Cosmic Egg are ever so slightly brighter than their surroundings. It is believed that these faint patches represent the large-scale structures we know today in embryonic form. To cosmologists, it was an unbelievably exciting discovery.

The picture will get better. Better "cameras", with improved technology, on soon to be launched satellites, will give us much more detailed pictures in the years to come. We will see more detail as the pictures will not be as blurred.

Ultimately, the detail in the photograph will be limited. It is after all, the picture of a gaseous surface, not the surface of a solid. Like the surface of a cloud, or the surface of our Sun, we can look in to a limited depth – and that blurs the vision somewhat.

But the discovery of the Cosmic Microwave Background is one of the greatest discoveries of all time – the moment when the human race first saw to the limit of the entire visible Universe, and when it realised it was literally contained within a Cosmic Egg!

Chapter Three

The Expanding Universe

Imagine for a moment that the Earth is a balloon gradually being inflated. The bigger it grows, and the more it stretches, the further things move apart. The airfare between London and Los Angeles is forever going up, because the distance between those two cities keeps increasing. A long distance runner, attempting to cross America, travels endlessly without ever reaching his destination. Oceans grow ever wider.

Well, airfares do gradually go up, and the Atlantic Ocean at least is getting wider (by a few centimetres a year), but not because the Earth grows bigger. Its size is fixed. But not the Universe in which it lies. That, as we shall see below, is always growing bigger – and has done so since its beginning.

Our Universe, in its present form at least, has a *finite* age. This formed the backbone of the discussion in the previous chapter, where the age of the Universe dictated the existence and radius of the *Cosmic Egg*. We look out to a limiting distance, back to an early epoch, shortly after the beginning. But why do we believe in the finite age? We have postponed discussion until now.

We find ourselves installed in a Universe, with everything up and running. The stars are shining. The most essential requirement for our habitation is the availability of heat and light from one of those stars – our Sun. Our existence is dependent upon it. It has kept our planet neither too hot, nor too cold, so that abundant water exists in liquid form. It has held it that way for billions of years allowing life to evolve and eventually the civilised state of modern mankind to arise.

Plain common sense tells us that, for the Sun to go on shining, it has to run on *fuel*. In the past, science sought the nature of that fuel and whether the energy extracted from it was gravitational, chemical or nuclear. Today we know that the fuel for the Sun is hydrogen, and that nuclear fusion, whereby it is slowly but continually converted into helium, is the way it operates. It keeps the Sun generating energy. It keeps the Sun shining. It keeps all the other stars shining.

Not only was I born with stars shining, but also here on Earth, *cars* were up and running. The analogy is a useful one, because cars, like stars, run on fuel. But cars only started running about a hundred years ago, and they are only likely to be around for another hundred years or so – at least the petrol-driven variety.

Why is this so? Simply because the planet has only a limited supply of fuel. Our cars and planes depend on a supply of crude oil (and the time-scale for replenishing that supply by natural conditions is hundreds of millions of years). All the oil in the world (in the planet at least) will only run the current number of cars for a *finite* number of years. Even if the planet were just one big fuel tank, sooner or later the Earth's population of cars would consume it. So the cars, that consume fuel, could not have been running for an infinitely long time, nor can they run indefinitely in the future. We might as well enjoy our motoring while we can.

So too the stars. They too consume fuel. They too could not have been running for an infinitely long time, nor can they run indefinitely in the future. The Universe described in the opening chapter is full of galaxies; each shining by the light of a million million or so stars. Consequently, the Universe in its present state could not have existed forever in the past, nor indeed can it exist into the distant future. Suffice for now to recognise that the evolution of the Universe includes a certain period when the stars are shining. This is it.

The only way around such an energy "crisis" would be a fuel re-supply system, but then the products of the fuel burning would also have to be removed before they filled up the entire Universe. Altogether, far too contrived to consider further.

Like a dying campfire, that burns brightly when fresh but subsides to glowing embers, the stars in our Universe will take a very long time to go out. But the evidence is that we are already well past the peak burning time when stars gave off their greatest amount of light.

So, even if somehow the Universe had always existed, the era of stars shining has a finite duration. The Universe, *as we know it now*, would still have a *finite age*.

A remarkable discovery, attributed to Edwin Hubble in the 1930s, shows the finite age of the Universe in a very convincing way. Hubble initially studied astronomy at the University of Chicago, but switched to law when he gained a Rhodes scholarship to Oxford. However, he returned to astronomy, and after seeing military service during the First World War, he took up an appointment at Mount Wilson Observatory, home of the then largest telescope in the world. By means of that telescope, Hubble was to contribute two enormous breakthroughs in the understanding of the Universe on a large scale. The first was the recognition of the proper status of galaxies (and about which much will be said in Chapter 4). From the second came an astonishing revelation: the Universe is *expanding.*

The expansion of the Universe is apparent by Hubble's realisation that, on a large-scale, galaxies are separating from one another. Outside our Local Group of galaxies (those galaxies closest to our own – as briefly described in Chapter 1), all galaxies are moving away from us. It is not that our Local Group is particularly unpopular; the same is true for whichever galaxy or group of galaxies you happen to live in. Like continents drifting apart, the galaxies are separating from one another. The galaxies do not get any larger, but the spaces between them are opening up (see Figure 3.1). The apparently empty spaces between galaxies are expanding. And so is the whole Universe.

Imagine a fruit cake baking in the oven. The currants in the cake are like the galaxies, the dough of the cake is like the space between the galaxies. As the cake bakes, it grows larger. The currants themselves do not grow larger, but they are pushed apart from one another as the dough expands.

Suppose we could somehow be situated on one of those currants, then we would see our nearest neighbouring currants move slowly away from us. If we looked to currants further away, we would see them moving away somewhat faster. If we looked to even further currants, then we would find them moving still faster. The more distant the currant, the faster it moves away from us.

That is what Hubble found with galaxies. Hubble had developed a technique

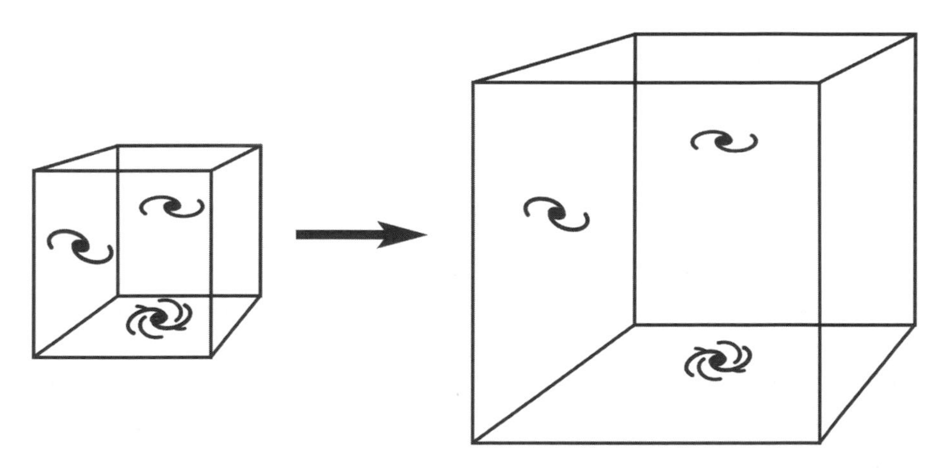

3.1 The expansion of space causes the separation between galaxies to increase, while the galaxies themselves do not grow larger.

for measuring the distances between galaxies, albeit relatively nearby galaxies by modern standards. He could put his information with that of Vesto Slipher, who had measured the "radial" velocities of galaxies, and found that those outside our Local Group were indeed moving away from us. Hubble showed that the more distant the galaxy, the faster it was moving away from us. He deduced a relation – that speed of recession is proportion to distance. Exactly like a fruit cake!

Vesto Slipher, an astronomer at Lowell Observatory, was the first person to measure the radial velocity of a galaxy – this is the speed of the galaxy directly toward or directly away from us. In fact, the first galaxy he measured was the Great Galaxy in Andromeda, and taking into account our motion within our own galaxy, he found it to be slowly approaching us. But once he chose galaxies outside our Local Group, he found the galaxies were moving away, many at several hundred kilometres per second. For this discovery he was given a standing ovation by the members of the American Astronomical Society at its meeting in August 1914.

The technique Slipher used is known in Physics as the *Doppler shift*. The principle is that light from any object moving away is slightly stretched in wavelenth. A similar principle is used by traffic policemen to operate radar speed traps. Doppler shift also affects sound waves, causing the sound of a receding vehicle to drop lower in frequency. Slipher dispersed the light of the galaxies into a spectrum, and then showed that the features were slightly stretched to longer wavelengths, that is to say toward the red end of the spectrum. The amount of stretching then allowed the speed of recession to be measured. Since the galaxies are almost always receding, their light is said to be *redshifted*. See Figure 3.2.

Edwin Hubble was able to measure the distances to galaxies by means of *Cepheid* stars – a recognisable type of star of known luminosity. Putting his results together with Slipher's velocities, he was able to see that the more distant the galaxy was from our own Milky Way, the greater was its velocity of recession.

For example, we see our nearest neighbouring galaxies (the ones outside our Local Group) moving from us at around a couple of hundred kilometres per second. Galaxies or groups at twice that distance move away at 400 kilometres per second, and the further we look out, the faster the galaxies are moving away. The Virgo Cluster, the central condensation of our local supercluster, recedes at 1,100 kilometres per second, those in the heart of the Centaurus Wall at 5,000 kilometres per second. Hubble's discovery has since been mapped to far greater distances – at least thirty times further than his original data penetrated, and, reassuringly, it still works.

It is tempting to think the galaxies were in flight from some central event,

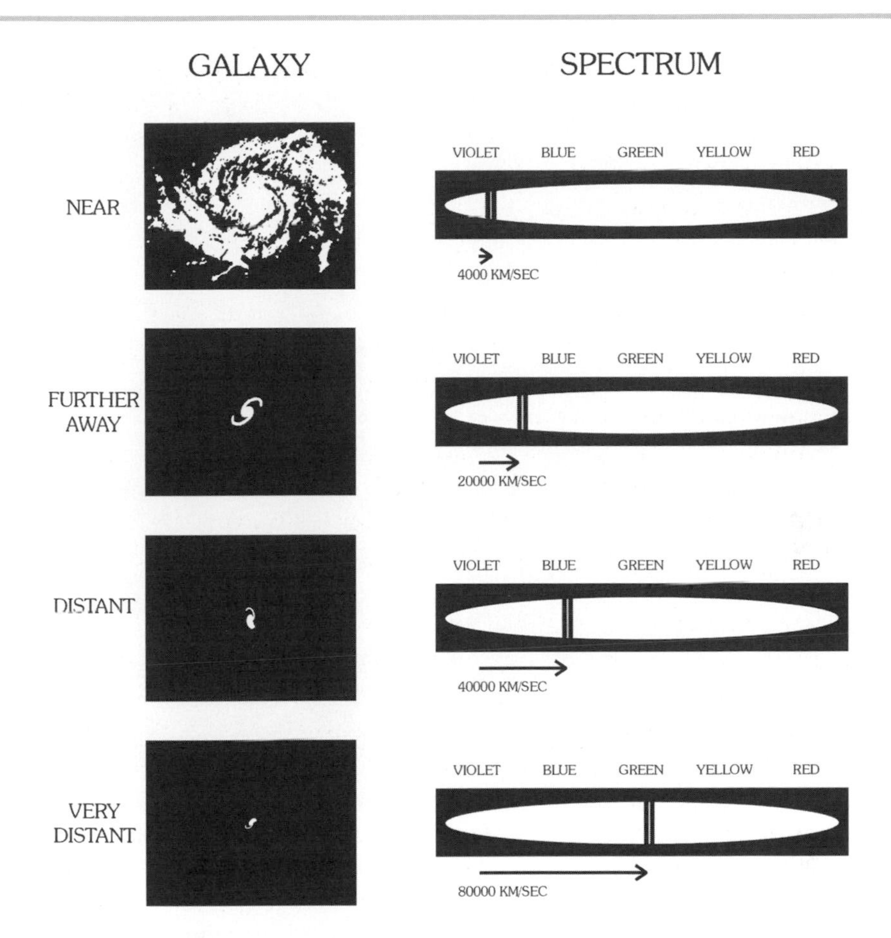

3.2 The more distant a galaxy, the faster its velocity of recession. The velocity can be measured by the "Doppler shift" in its spectrum of colours.

and that is how Hubble initially interpreted it. However, the galaxies are in fact as passive as the currants in the fruit cake. It's the cake – the space between the galaxies – that is growing, and thereby causing the action.

We are literally embedded in the cosmic fruit cake, and wherever we are in the fruit cake, the picture is the same; we think of ourselves as stationary, and see the neighbouring currants as moving away from us. This is illustrated in Figure 3.3, where we can either regard our own galaxy as stationary, with everything else in motion away from it – or any other galaxy as stationary with everything else, us included, in motion away from it.

Expansion is a fundamental property of the Universe. This is not easy to grasp, for we are inclined to believe that the Universe is a nice stable place. Few people question whether the stars will keep on shining, or whether they

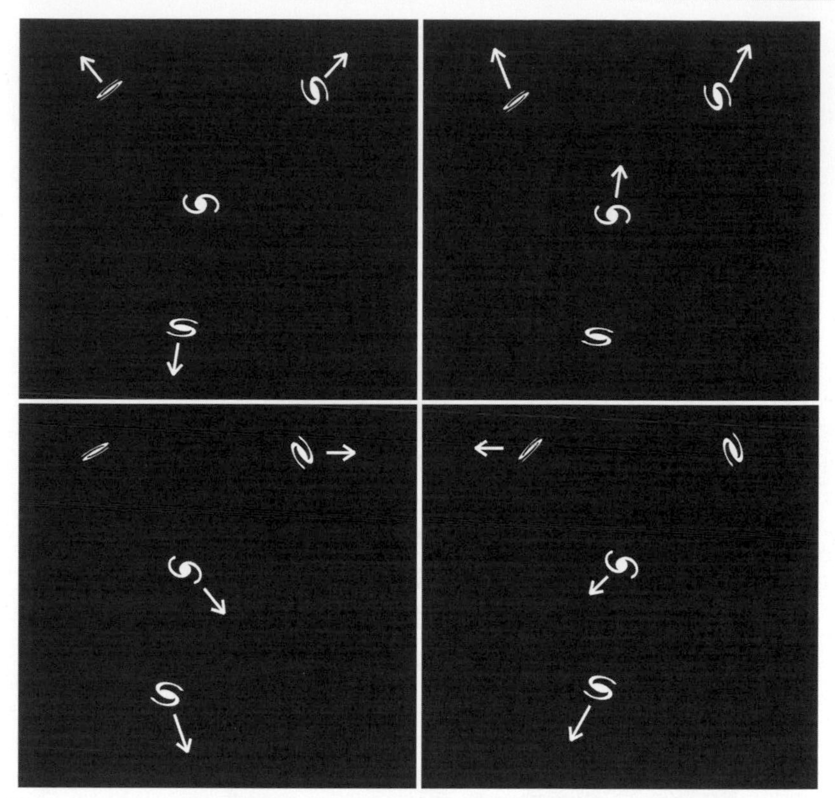

3.3 The expansion of the Universe appears the same regardless of which galaxy we think of as stationary.

have always shone. Similarly, the concept that we are riding a galaxy on a trajectory to goodness knows where is hardly a comfortable one.

So how does empty space manage to expand? After all, the cake uses baking powder – tiny granules that release gas when heated, each granule creating a tiny cavity in the mixture. But can empty space expand?

It would seem impossible in the classical *Newtonian* approach, but Einstein's *Relativity* has shown us that the basic dimensions of space are not as rigid as expected. They have an elasticity. They can distort, they can stretch, they can contract. It is possible for the dimensions of the entire Universe to expand, and they could equally well contract. (In Chapter 8, we will look at Einstein's approach in far more detail – for now, think of space as the dough in the cake.)

In fact, the mass of the galaxies embedded in the fabric of space will tend to cause space to contract. It's a matter of gravity, and even Newton would have galaxies attract one another. But, in Einstein's General Relativity, the galaxies are much more like the currants in the cake. If the currants were attracted to one another and tried moving closer to one another, then in the process they would cause the entire cake to collapse. Similarly, if the galaxies were not flying apart from one another, then they would pull themselves toward one another – and in the process, they would squeeze the space between them – and cause the whole Universe to collapse!

Einstein's revelation of the elastic properties of space makes the whole Universe behave like a stone thrown up in the air. So long as it is flying upwards, it resists the pull of gravity. But gravity gradually slows down the stone until it is momentarily at rest, then falls.

So too with the Universe. If there is enough gravity, it will eventually gain the upper hand, bring the expansion to a halt, and then cause the whole Universe to collapse – pulled in by the gravity of its galaxies. The cake simply collapses.

The fate of the Universe – expansion forever or eventual collapse – has long been the key question in cosmology. Enormous observational efforts have been made, but without going into details, no compelling evidence one way or the other was found. However, the current general concensus is that there does not seem to be quite enough mass in the Universe – in the galaxies or even in haloes that surround them – to arrest the expansion. The expansion will therefore likely go on forever, and thus the Universe will likely go on forever, but eventually there will be no fuel to light the stars and the Universe will never be the same again.

While we may debate the future, we can be fairly sure of the past. The expansion of the Universe opens a door for looking back in time. Suppose, as in Figure 3.1, one could document the expansion of the Universe with a sort of video camera that recorded the galaxies moving apart. Run the video backwards, and the galaxies appear to move closer. And, if somehow, one could keep going backwards in time, then all the galaxies would come together!

In other words, there had to be a time when it all began. The expansion has not gone on forever – or infinite distances would now separate the galaxies. The galaxies are separated by finite distances – so the age of expansion must be finite. The age of expansion is the age of the Universe – in its present form at least.

This overall scenario is known as the *Big Bang*. In fact, the name was given derisively by the famous cosmologist, Fred Hoyle, who had championed an alternative *Steady State* theory that had matter continuously created in the voids to form new galaxies. However, that theory has since been generally disproved and is no longer taken as a serious contender.

The term *Big Bang* correctly suggests an explosive beginning to the Universe in its present form. But this brings about a misconception. The explosive beginning is all too easily likened to a bomb exploding in otherwise empty space – the galaxies flying outward from a central point. That is

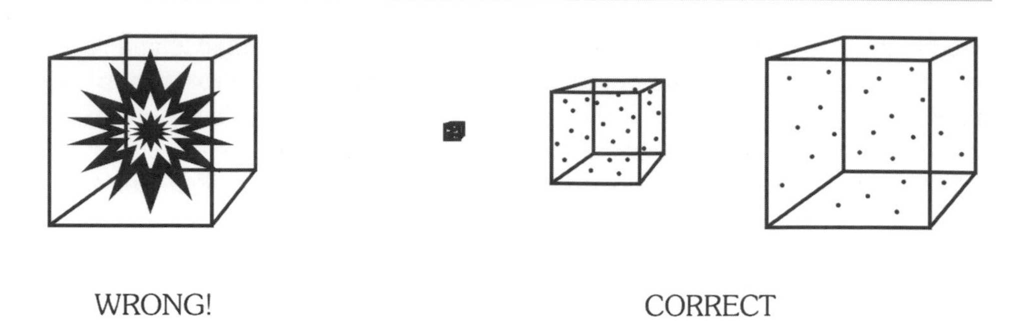

3.4 The big bang is not like a bomb exploding in an empty hall, but rather like the "hall" growing explosively larger.

incorrect! Rather it is *space that grows explosively*! And the matter within it is simply carried along. As suggested in the figure, it is not that the matter is flung out by the debris of the explosion, rather that the Universe is already filled with matter, and matter always fills the volume as the Universe expands.

Not only did our Universe begin with a Big Bang, it began with a *Hot* Big Bang. Expansion causes cooling – a fundamental fact of Physics – so the Universe is both expanding *and getting cooler.* In the past, the Universe was denser, and warmer. If we could travel backward in time, to a period soon after the Big Bang, then the Universe would be very much denser and very much hotter.

In its first instant of existence, the Universe would be unimaginably hot, and unimaginably dense. We cannot, of course, see back to this time, but our knowledge of Physics has given us remarkable insight into those early moments of the Universe. Some years ago, the now celebrated physicist Steven Weinburg, captured the essence of that early beginning in a classic book entitled *The First Three Minutes.* It deals with that fleeting but crucial era when the temperatures and densities were so high that exotic high-energy nuclear processes caused matter to be created from radiation, and radiation to be created from the annihilation of matter. The rapid cooling led to a sorting out of nuclear particles as conditions became more like those that exist fleetingly in the ignition of a hydrogen bomb. In the ensuing mêlée, the matter content of the Universe was turned into 76 percent hydrogen and 24 percent helium, and virtually nothing else. Hydrogen was much, much later to play its role as the main fuel source for the stars, but the issue as to how elements heavier than hydrogen and helium were created is to be taken up later.

And so, the early content of the Universe was a hydrogen/helium mixture kept in gaseous form by its high temperature. Not unlike the present interior of the Sun – but filling the entire Universe. Thus it set the stage, described in the previous chapter, whereby the Universe remained in such a state of incandescence until it aged some 300,000 years, and had expanded and cooled enough for its magical transformation from an opaque mass into a kingdom of transparency.

The uniform distribution of material throughout the Universe did not endure – for the Universe today is a very different place. Today, matter is concentrated into galaxies, within which much is further concentrated into stars etc – as described in the opening chapter. It is just as well, or we could not be here, or even exist in a Universe without stars or planets.

Gravity is believed to have brought about that transformation. The slightest over and under-densities provided the seeds – the over-densities had slightly more gravity, the under-densities had very slightly less. Matter was very gradually drawn away from the under-dense regions toward the over-dense regions, thereby exacerbating the dissimilarity. The greater the difference, the greater the flow of matter toward the higher density regions. Eventually almost all the matter was to collect into galaxies and large-scale structures. The picture seems straightforward, but (as we shall read in chapters ahead) there are complications.

The general scenario, however, seems complete. From a Universe born in a fiery big bang, large-chunks of matter condensed, eventually to form the galaxies of today.

So far, so good, the essential information of this chapter has now been covered. However, from here to the end of the chapter, the going becomes a bit heavy, to say the least. Some readers may wish to opt out, and jump to the start of the next chapter.

A paradox has been presented. From where we live we can look in two opposite directions – say 15 billion light years to the left and 15 billion light years to the right, right out to the origin of the *Cosmic Microwave Background* – the shell of the Cosmic Egg (in the previous chapter). And in both directions *it looks the same*. Why is this a paradox? Well, we are seeing two different parts of the Universe that are now separated by 30 billion light years. Those two regions have never had contact with one another – each has always lain *outside the other's Cosmic Egg*. During the history of the Universe, they have had nothing to do with one another. Why then do they look so similar?

We have already seen that the Cosmic Microwave Background gives us an image of the early Universe, when it was no more than 300,000 years old – by cosmological standards, in its infancy. Yet that Cosmic Microwave Background is incredibly uniform. Its average brightness on one side of the sky matches that on the opposite side within one part in a million. Furthermore, the famous fluctuations discovered in the background are at most a few millionths above or below this level. The Cosmic Microwave Background is so similar on one side of the sky to the other, yet those pieces in opposite directions are so far apart that each is beyond the "horizon" of the other. They could never have had any contact.

Similarly, the Hubble Space Telescope has taken two extremely deep photographs (described further in Chapter 5) in opposite directions. The photographs are crowded with galaxies, many some ten billion light years out, or further. But the two photographs look very similar – even though the galaxies on one side lie beyond the "horizon" of those on the other side.

How can such *uniformity* on such a large scale come about? We have no observational evidence to answer the paradox. But in 1980, physics seemed to come up with a theoretical solution. *Inflation.*

The "horizon" problem could be overcome if only the Universe could be made to expand very rapidly and then slow down. In other words a period of incredibly rapid expansion should precede the present somewhat sedate expansion. The *inflationary* expansion would come in the first fleeting moments of the Universe's existence. The subsequent expansion – first inflationary, then conventional – would be so enormous, that what started off about the size of an atom would become the size of a chicken egg, that would in turn grow to the size of the entire Universe now visible inside the Cosmic Egg of today!

Not only does it seem to stretch space to its limit, it does the same to our imagination! Perhaps it helps to realise that the present world is dominated by emptiness – not only the almost empty spaces between planets, stars and galaxies, but even the emptiness of the atoms of which they are composed (as will be discussed further in Chapter 9). The density of everything when the Universe was so tiny must have been incredibly high, but to physics, it was possible.

What drove the inflationary expansion? Again an answer has been forthcoming from physics – but at rather a technical level. For those who know something of physics at an introductory level, it involves an equivalent of *latent heat.* The energy to drive the inflationary expansion came about from a phase change in the very very early Universe. It is something with aesthetic appeal to the researchers, that involves *field theory,* that has been highly successful in the understanding of the behavior of matter on the nuclear scale. It can be applied to the entire Universe at a time when it existed at a nuclear density.

Can we have confidence in using physics to extrapolate back to the very very early Universe, involving bizarre conditions and circumstances – like inflation? It is difficult to say, but all good theories of Physics make predictions, and those predictions can be tested today. The theory behind Inflation not only solved the horizon problem and why the Universe looks so uniform today, but also predicted that the Universe today should have a density today almost exactly on the *critical density*. This is the density of matter just sufficient to arrest the expansion of the Universe.

Initially that prediction seemed to hold. In the 1980s, the recognition of there being substantial amounts of *dark matter* in the Universe, as well as the luminous galaxies, was widely accepted. However, by the 1990s, various tests had been devised that reported that the density of matter was only about a quarter or a third of what was necessary. Some have seen it as a major problem for the inflationary theory, others as an opening for modifications to inflation, particularly in the light of the possible need for *antigravity* – the subject of discussion of a later chapter. The lack of matter may be compensated by a contribution from antigravity.

Nevertheless, whether or not inflation is real does not affect the reality of the expansion that we see today – the fundamental property of the Universe, against which everything else in cosmology takes place.

For now let us return to the realities of observation – the reality that has us living within that apparent hollow shell, the Cosmic Egg.

As depicted in Figure 3.5, the Egg grows larger as the Universe expands. Its radius grows at the speed of light. Since the inside of the shell is receding from us at very nearly that speed, the contents of the Egg are hardly seen to change over time. However, the stars and interstellar matter within the galaxies slowly evolve, and while the Egg expands, the galaxies themselves do not grow larger.

The expansion of the Universe not only causes the Cosmic Egg to grow at the speed of light, but whatever is outside the Egg grows faster still. As shown

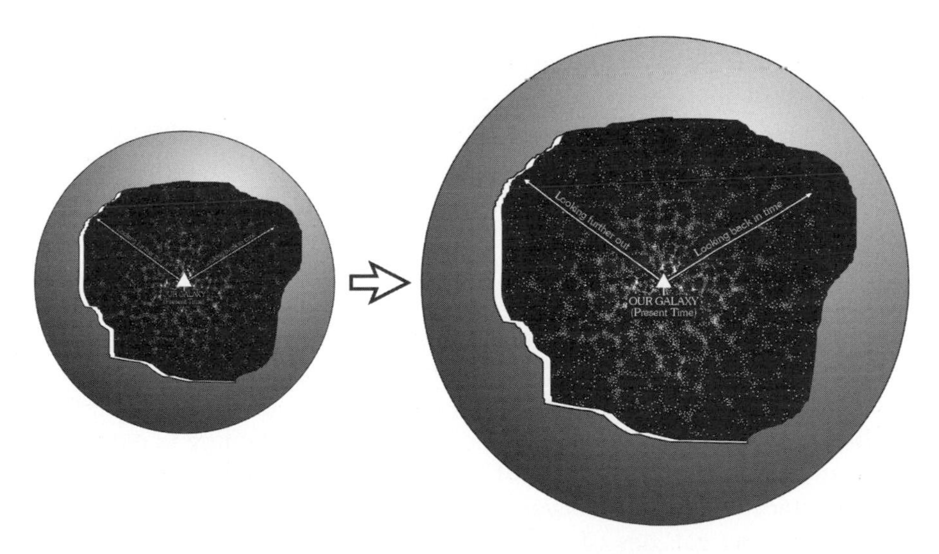

3.5 The Cosmic Egg grows bigger as the Universe expands.

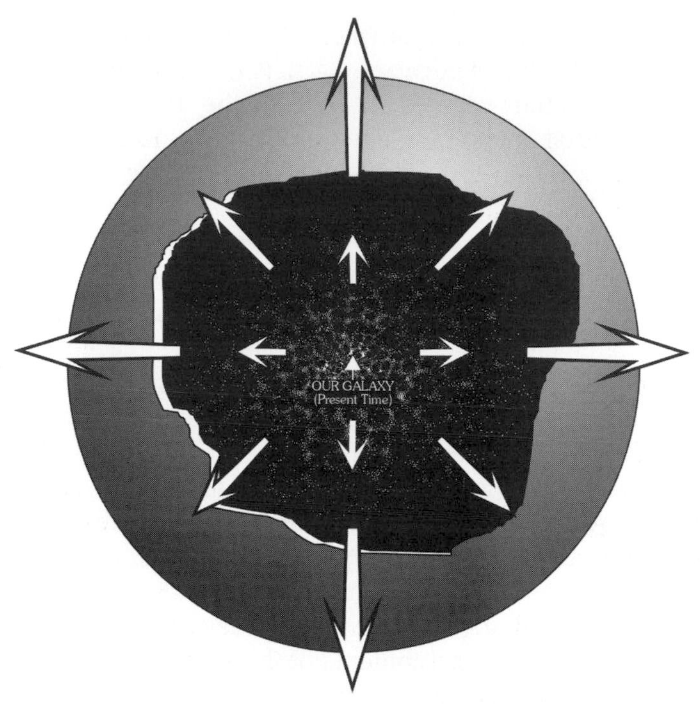

3.6 The space inside the Cosmic Egg, its shell, and the space outside the Egg, are expanding with ever greater velocities from our position in the centre.

in Figure 3.6, we can set our own Galaxy as stationary in the centre of the Egg. Then, as we see, the velocity of expansion increases with distance from that central point. The greater the distance, the greater the velocity. When the inside surface of the Cosmic Egg is reached, the velocity is very close to the speed of light.

Just outside that inside surface we can mark a sphere, where the expansion velocity is the speed of light. In effect, this is the outer shell of the Cosmic Egg. Had the shell been transparent – but it isn't – this would have been the limit to the visible Universe.

Outside the Cosmic Egg, the speed of expansion becomes still greater and greater with distance from the central point. In other words, greater than the speed of light. Impossible? Surely, this violates the basic foundation of physics. Einstein proved that nothing could travel faster than the speed of light. Not quite. It is acceptable for the Universe outside the Cosmic Egg to expand away from us at greater than the speed of light, because we cannot see it, and we have no contact with it whatsoever.

In any case, the galaxies (if there are any) that are exceeding the speed of light are only doing so *relative* to our Galaxy. They themselves may be quite stationary relative to their surroundings. It is space itself that is doing the expansion, space itself that is moving greater than the speed of light, relative to us.

And the roles could always be reversed. If there is a galaxy currently in existence outside the Egg, then, were it to be inhabited, the view that those inhabitants get of the surrounding Universe would be their own Cosmic Egg. It would look just like ours, but it would have a different centre. And, relative to that centre, we would be travelling at faster than the speed of light. Remember too that the Cosmic Egg is our *vision* of the Universe – it is not a physical structure.

Time to slow down. No, not the Universe, just our presentation. We have covered all the essential aspects of the expansion of the Universe, the overriding motion that affects all of the large-scale aspects of the cosmos. To readers who have not previously known or understood it, the going has undoubtedly not been easy. The expansion normally proves to be one of the most baffling properties of the cosmos – particularly how empty space can just expand.

In such circumstances, analogies are much easier. Think of the fruit cake baking. It is a much easier concept, and it tells you all you need to know about the expansion of the Universe.

Chapter Four

The Cosmic Labyrinth

The Universe may be full of billions of galaxies, but only *four* are visible to the naked eye!

One is our own Galaxy, seen in the night sky as the Milky Way. The other three (also described earlier in Chapter Two) are the two Magellanic Clouds and the Great Galaxy in Andromeda.

To see further galaxies, some optical aid is necessary, and even a pair of binoculars makes a big difference. They let you find most of the objects in the *Messier Catalogue* – a list of some one hundred *nebulae* drawn up in the 18th century by a Frenchman, Charl Messier. It is still in popular use today.

Nebulae – Latin for *clouds* – are fuzzy patches of light seen in the sky, in contrast to those pinpoints of light that are the stars. They are relatively rare. Your pair of binoculars is probably capable of seeing up to fifty thousand stars in the sky, but only a hundred or so nebulae.

Not all nebulae are galaxies. Some are gas clouds within our Milky Way, or very distant clusters of stars. In Messier's catalogue of some one hundred objects, only about thirty are galaxies. However, in general, nebulae seen against the Milky Way are objects in our Galaxy; nebulae seen in the rest of the sky, clear of the Milky Way, are almost invariably galaxies. It is the usual problem in astronomy – small things nearby mimicking large objects far off.

A number of Messier's nebulae lay in a small part of the sky – the constellation of Virgo – well away from the plane of the Milky Way. We know in retrospect that all these objects were galaxies, and that the Virgo Cluster is the most prominent nearby concentration of galaxies.

The late 18th and early 19th centuries were to see a revolution in observational astronomy, carried out by the father and son combination of

William and John Herschel. Though born in Hanover, William Herschel spent most of his life in England. Originally a successful musician and composer, he turned to astronomy as a hobby. He was soon making telescopes superior to any others, and using them too. His discovery of the planet Uranus – the first beyond the "big five" visible to the naked eye – led to fame and a pension from the king (also Elector of Hanover) that allowed him to devote his energy full time to astronomy. His further discoveries included numerous double stars, moons of Uranus, infrared radiation, and some 2,500 nebulae!

His only son – John – was also to make enormous contributions to chemistry, photography (his word), philosophy and botany. But his greatest work lay in following his father's footsteps. In particular, he resurveyed all of his father's nebulae, added some five hundred more, and then relocated his telescope to the southern hemisphere (to the Cape of Good Hope – just a couple of kilometres from where this book is being written). The final result was the General Catalogue of some 4,600 nebulae, 90 percent of which were first seen by either William or John Herschel. Given that galaxies formed the great majority, the catalogue provided the first revelation of the character of the Universe on a large scale (see Figure 4.1).

It was some time before the Herschels (and other astronomers) understood what they had observed. Things might have been simpler had William stuck to his original belief – and John to his own belief – that *most of the nebulae were galaxies*. William had correctly surmised that many of the nebulae were stellar systems, too distant for individual stars to be resolved. This married with the then current philosophical concept of *island universes*. However, later in life, William changed his mind, being swayed by the alternative view that all nebulae were simply gas clouds in our Galaxy, possibly in the process of forming new stars.

The overriding problem was that there was no way to measure the distances to the galaxies – that had to wait until the 1920s. So scientific opinion was polarised between those who believed that all nebulae were part and parcel of our stellar system, and those who (correctly) regarded most of the nebulae to be independent stellar systems, too distant for their stars to be seen individually.

John Herschel believed the latter – though he has later frustrated historians by seeming to support both sides. The truth is he revered his father's opinion, and also was strongly criticised for his island universe interpretation. Realising he had no convincing evidence, he therefore went along with the more conservative alternative – gas clouds in the Galaxy – in the many versions of his classic textbook *Outlines of Astronomy*.

However, having made by far the most complete survey of the nebulae – and deduction. He saw that a third of all nebulae were concentrated into an eighth

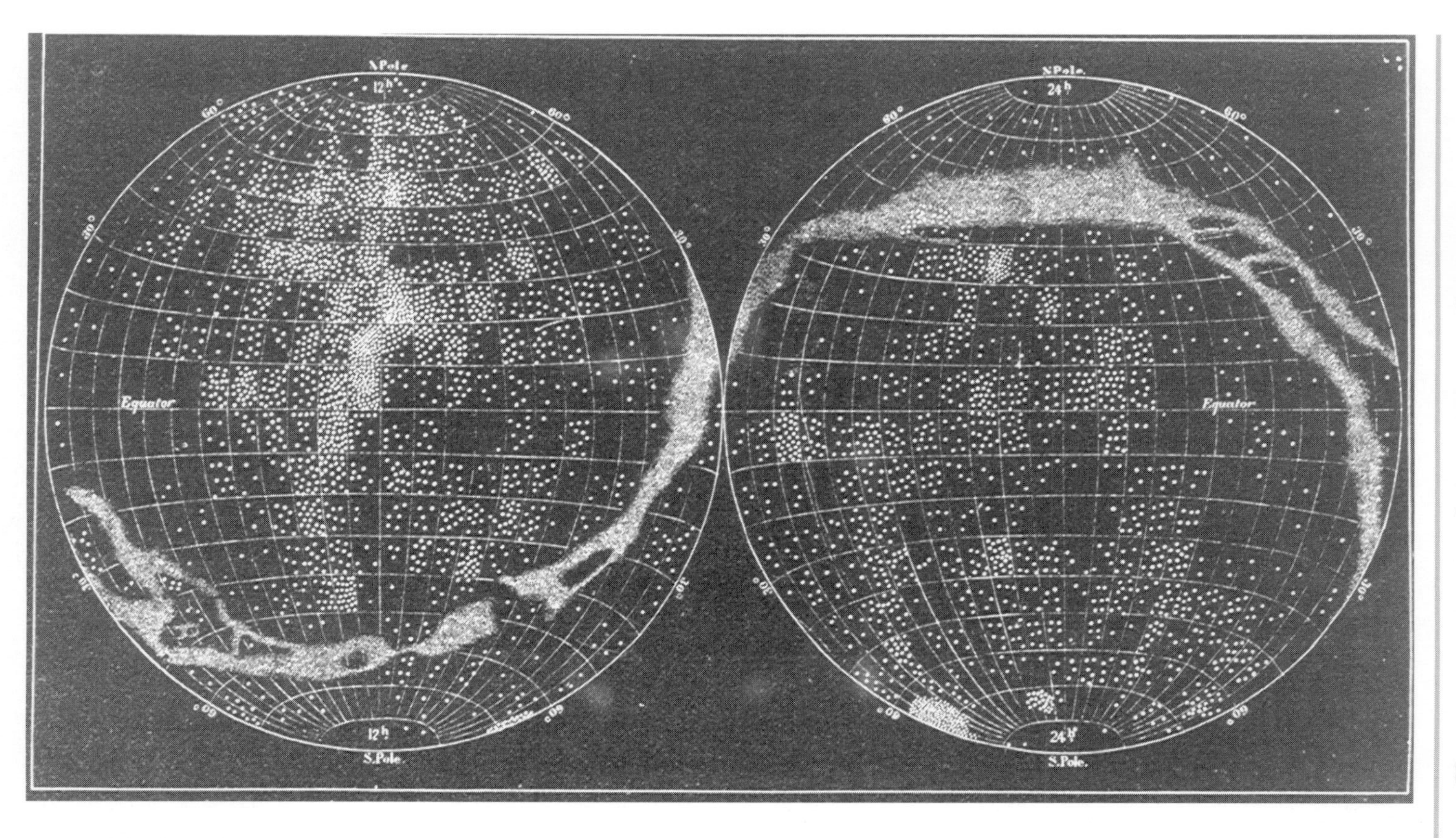

4.1 *This plot by Richard Proctor, published in 1878, shows the distribution of Herschel's nebulae over the entire sky. The Milky Way is obvious. Note the great concentration toward the constellation of Virgo, in the centre of the left-hand panel. (From R.A. Proctor,* The Universe of Stars, *published by Longmans, Green, and Co., London.)*

being able, for the first time, to see how they were spread over the whole sky, he made an astonishingly correct of the sky – that centred on the constellation of Virgo. He therefore supposed that our own Galaxy lay far from this region but was "involved within its outlying members". He saw an irregular distribution, with branches – or "protuberances" – running outwards from the core in Virgo, and deduced that our Galaxy "forms an element of some one of its protuberances". In so doing, he was the first to discover *large-scale structures* in the Universe.

He could have gone even further. In subsequent sky plots of Herschel's *General Catalogue* of Nebulae by Richard Proctor (see Figures 4.2), one sees

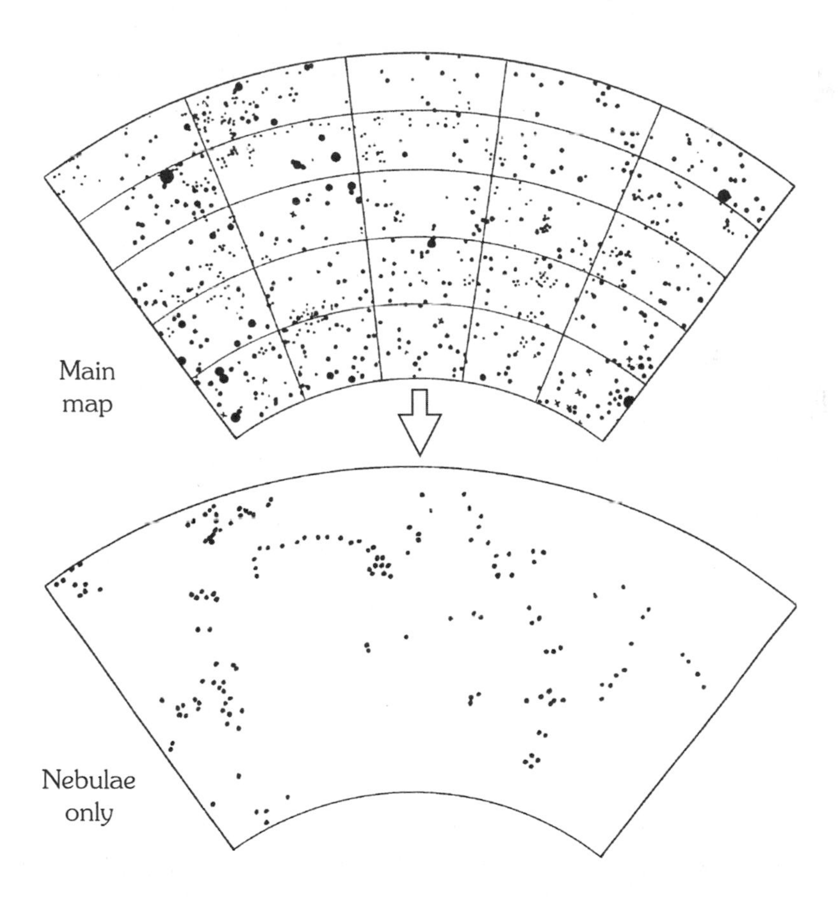

4.2 A portion of a plot by Richard Proctor shows both stars and nebulae (upper panel). If, however, only the nebulae were plotted (lower panel), then their extraordinary distribution would be apparent. (From R.A. Proctor, The Universe of Stars, *published by Longmans, Green, and Co., London.)*

that often the nebulae assemble into filaments surrounding otherwise empty voids – the first sighting of the cosmic labyrinth. But Proctor, a populariser of astronomy in Victorian England, strongly criticised Herschel's interpretation, and did his best to sway readers to the more conservative gas-clouds-in-the-Galaxy hypothesis. Indeed, some were – as shown by spectroscopic observations – but these were the nebulae near the plane of the Milky Way.

The *great debate* (see Figure 4.3), as it became known, came to a head in the early twentieth century, with the advent of improved photographic emulsions and the new *100-inch telescope* at Mount Wilson Observatory, high above Los Angeles. Using the new telescope, astronomers no longer saw nebulae as blurs. Some clarified into elegant spiral structures, and the telescope could photograph them by the thousand.

Through the great telescope, the closest of the spirals (the Great Galaxy in Andromeda) appeared to resolve into pinpoints of light, but it took the genius of Edwin Hubble to identify particular *Cepheid* stars amongst them. In doing so, he proved they were really stellar systems in their own right, and the Cepheids acted as distance indicators. He put the distance to this nebula as some 500,000 light years, and although this estimate is only a quarter of the true distance known today, it was clearly enough to establish that most of the nebulae were indeed island universes, or galaxies.

In the years that followed, Hubble was to combine his distance estimates with the radial velocities of Vesto Slipher and thereby produce the relationship that showed the Universe to be expanding (as was recounted in the previous chapter). He therefore made two monumental contributions to the opening up of the extragalactic Universe, and his classic textbook *The Realm of the Nebulae* summarises his work.

Hubble was also one of the first to use photography to look at the *distribution of the nebulae*. His survey covered over a thousand fields, but each field of the 100-inch Mount Wilson telescope was small in extent while probing relatively deep. Consequently, he made and promoted the interpretation that the nebulae were distributed, much like stars, in more or less a uniform but random fashion over the entire sky, save for a very few rich clusters and an absence where they were obscured by the Milky Way (see Figure 4.4). Clearly, Hubble had never read Herschel's description of the great system centred in Virgo.

In fact Hubble pushed this line to contradict his protagonist, Harlow Shapley. Shapley had also been on the staff of Mount Wilson Observatory, and had backed the wrong side in the *great debate*, before securing the directorship of Harvard Observatory. He is best known for having shown that our Solar System was far removed from the centre of the Galaxy. In the 1930s, he had claimed the existence of *Metagalaxies* – clouds of very distant galaxies – in the Southern sky, recognised on wide-field photographs. It was Shapley who also

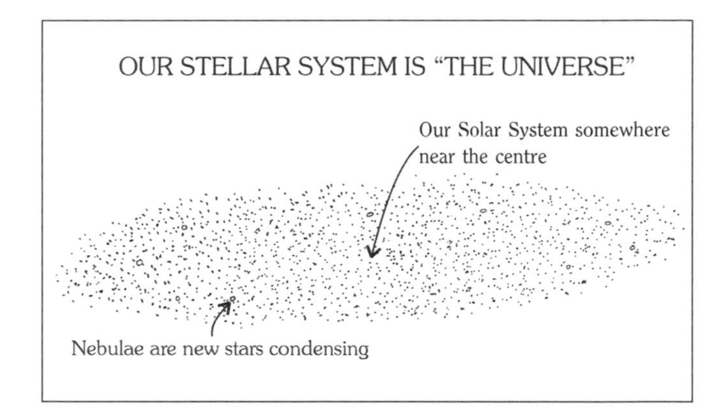

OR

4.3 The "great debate" as to whether the cosmos was populated by a single stellar system, or by multiple systems (known as island universes) spanned the 18th to 20th centuries.

advocated the use of the word *Galaxy* to discriminate between the true "external" or "extragalactic" nebulae and the nebulae within our own Galaxy. Again, Hubble did not go along with this, and insisted on retaining the traditional word *nebulae.*

Today we know that, in terms of the distribution of galaxies, Hubble was wrong, and Herschel and Shapley were right. Great superclusters of galaxies

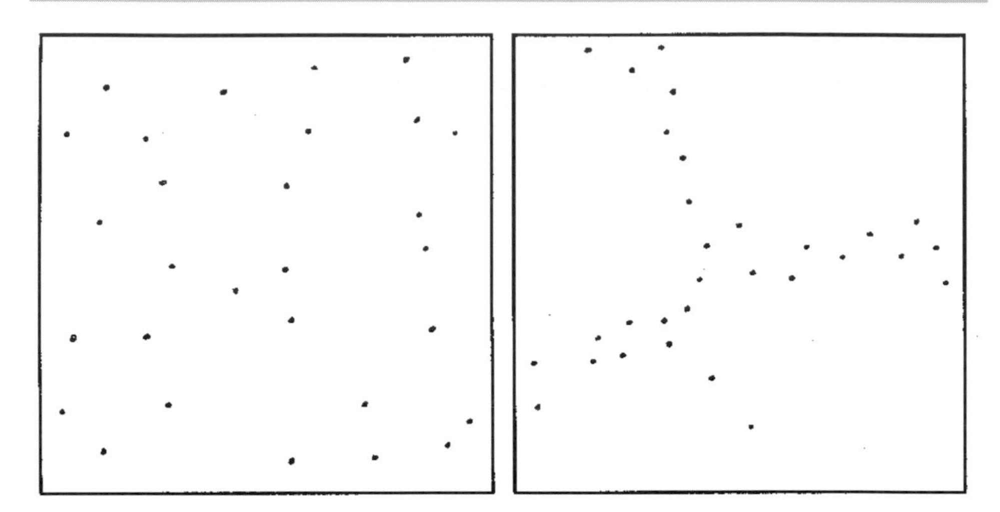

4.4 The distribution of nebulae (galaxies). Hubble believed the nebulae were scattered randomly (left), whereas we now know that is far from the case (right).

do exist. Their shapes are irregular – much as John Herschel described in the "protuberances" extending out from Virgo in our own supercluster. But, surprisingly, superclusters are not separate entities in themselves; rather they interconnect to form the great *cosmic labyrinth*.

The recognition of superclusters in the late twentieth century was a vindication of a few lone scientists who had championed their existence. However, the evidence first apparent in the Herschel/Proctor maps – that there were empty voids surrounded by galaxies – also resurfaced.

There was one great advantage – Hubble's discovery that the *redshift* of a galaxy was (thanks to the expanding Universe) an indication of its *distance*. Knowing distances, we could not only plot where the galaxies lay in the sky, but could also construct a *three-dimensional* map of where the galaxies lay in space. Moreover, the introduction (in the 1960s) of electronic image intensifiers suddenly meant that telescopes could measure redshifts far more rapidly. Whereas exposures of many hours used to be required to record a galaxy's spectrum onto photographic emulsion, the new detectors were ten times faster, and so the number of available redshifts – and distances – increased dramatically.

By the late 1970s, Jaan Einasto, an astronomer who led a group in Estonia (then part of the Soviet Union), made use of a compilation of redshifts within a newly published catalogue. At an international conference which he hosted, he and a colleague presented a paper that asked if the Universe had a *cellular structure*. He saw *voids surrounded by walls of galaxies!*

It was a sensational claim and the news travelled far and wide, but few believed it for it was so totally unexpected. Nature didn't do that sort of thing – well, not on that scale. The laws of physics never predicted it. "Surely, it couldn't be" was the consensus a colleague and I reached, after an afternoon coffee discussion in Nashville.

The chilly political climate did not help. American astronomers, in particular, didn't want to believe it. It seemed almost contrived to support the "pancaking" theory of gravitation collapse that the Russian theoretician Zeldovich was pushing at the time. Furthermore, an eminent and highly respected group of American researchers were already pursuing an alternative theory, which they thought had already successfully predicted the character of the distribution of galaxies – and it was not cellular in any way. Perhaps the most biting criticism that could be levelled at Einasto's work was that it was based on uncontrolled *catalogue* data, and not on a proper statistically controlled data set.

But the truth will out, and in the years that followed, American researchers mapped superclusters and voids. The number of catalogued redshifts grew, and plots confirmed a cellular texture to the way galaxies were distributed in space. Formal confirmation came from two redshift surveys (using statistically controlled data) from the *Harvard-Smithsonian Center for Astrophysics*.

It also proved me wrong. When I was a student, I regretted not being around fifty or more years earlier, when Hubble was about to make his great discoveries. It seemed to me that by the time I came along, almost everything was understood and only the details needed filling in. Yet, here was a sensational revelation as to the character of the Universe, and it only came out in the 1980s! Fortunately, I too have played a part in the mapping of the cosmic texture, particularly in the southern skies, and it is something close to my heart.

In the remainder of this chapter, we will explore the character and local features of that cosmic texture. In the diagrams, we will see the galaxies represented as points, as though they were little particles in space. It is as well to remember that each little point is really a great city of stars! We will also talk of distances between galaxies in terms of millions of light years, as though a few million means just next door, while remembering what vast separations they really are.

The essence of the cosmic texture can best be seen in Maps 5a and 5b, in the second colour section of this book. Many have described it as frothy and foam-like, as though it would be more at home in the head of a tankard of beer, than in the cosmos. It suggests that galaxies, or the material from which they are made, behave more like particles in a liquid than particles that move freely from one another. Particles in a *soapy* liquid tend to adhere to one another, instead of breaking up into drops. Blow air and the liquid holds itself together, wrapping around the bubbles. Could this also account for the cellular texture of the cosmos?

Galaxies do tend to *adhere* to one another. They do not like to exist in isolation. Almost all galaxies have at least one neighbouring Galaxy, never more than 10 million light years away. Given this tendency, they assemble into gatherings, like disorderly soldiers assembling into platoons, and the congregations they form are called *large-scale structures*. These structures are coherent and have fairly irregular shapes – the *superclusters*.

There is a tendency for the superclusters to show overall flattened forms – usually referred to as *great walls*. Alternatively, some show flattened but elongated forms, more appropriately termed *great ribbons*. Neither walls nor ribbons are particular thin, usually having a thickness of some 50 million lights years – whereas their major dimensions may run toward 500 million light years or more.

It is also difficult to say where one structure ends and the next begins, since all large-scale structures appear interconnected to form a labyrinth. This great cosmic labyrinth looks much more like something contrived in the cyberspace of a computer than reality. But reality it is. Had we the means to travel, we could journey from our Galaxy to any other Galaxy, via galaxies that were never more than 10 million light years apart – a bit like driving to any other town, where we could always expect to fill up at intermediate towns never more than a certain distance apart. Of course, our path might be far removed from a straight line, but more like some game, we could always interconnect any two galaxies in the Universe by an indirect route that jumps from galaxy to galaxy.

Some large-scale structures appear more dominant than others, and often light frothy-like structure may form the interconnecting fabric. Within the network of the more dominant large-scale structures, occasional clusters of galaxies occur. Here the galaxy to galaxy separations are down to less than a million light years – a very much denser concentration of galaxies. Often, these clusters tend to form central condensations to surrounding superclusters and sometimes occur at the confluence of major structures.

Large-scale structures account for only some ten to fifteen percent of the volume of the Universe. The rest – the spaces between the labyrinth – is apparently empty. Voids, ranging up to 300 million light years across, are packed like inflated balloons to cushion the texture of the cosmos. Like the balloons, the voids have a tendency to be spherical – at least their widths, breadths and heights are similar. They come in all sizes – down to the scale where they lose their definition amongst the almost point-like galaxies.

They give the Universe that frothy texture, which is more like the texture of a bath sponge than beer foam. The material from which the sponge is made is like the large-scale structures. It is all interconnected – otherwise pieces would be forever falling out of the sponge – or bits would rattle around inside when the sponge was shaken. The bubbles in the sponge are like the voids between the large-scale structures, and they too are all interconnected – so that water can be absorbed into, and squeezed out, of the sponge. So, too, there seem to be sizeable openings linking every cosmic void.

One might chose to "travel" from any galaxy to another (not necessarily in a straight line), by keeping only to the high-density regions of the Universe, and never passing through a void. Alternatively, one might rather chose to travel around the Universe keeping well clear of any galaxy, using the network of voids.

Voids are ubiquitous – you find them everywhere – like cosmological baking powder, the analogy used in the previous chapter. I cannot claim to have examined baking powder under a microscope, but I suppose its granules range in size. Bigger granules give off more carbon dioxide when heated, so they create bigger bubbles in the cake, and smaller granules create smaller bubbles.

So too with the cosmological voids. It is as though between the large-scale structures, bigger granules had been used to make bigger voids. Yet smaller voids seem to percolate the irregular forms of the large-scale structures – as though smaller granules had been used there.

One could go as far as to say that a fairly accurate description of the cosmological fabric is that of a cake with a bubbly effervescent texture, but where the performance of the grains of baking powder has varied from region to region. Perhaps this provides a considerable clue as to how the cosmos evolved. (We will defer much of this discussion to a later chapter.)

Where then do we find ourselves – in our Galaxy – in the cosmological cake? Figure 4.5 shows the layout of the local texture. "Local" is understood here to range out to a few hundred million light years. Learning what superclusters are situated in this region is a bit like learning what continents there are on Earth. These are the "continents" of the extragalactic realm, separated by oceans of space. Let us take a tour.

To begin, Sir John Herschel's description of the most local of textures is quite accurate, for we are now known to be a part of the *Local Supercluster* or *Virgo Supercluster,* centred on the Virgo Cluster (named for the foreground constellation of stars). Various extensions – arms or protuberances – seem to run roughly radially outward from this central condensation, and our Galaxy is in one of them. The supercluster has quite a flattened form. There are also many voids – between the protuberances, and especially above and below the supercluster. That "below" is known as the *Local Void.* Our Galaxy lies close to it. Save for a couple of foreground galaxies, we can stare straight into nothingness. (Aside from the diagrams in this chapter, the Virgo Supercluster, and other local features described below, can also be seen in the colour *Universe in 3D* section, elsewhere in this book.)

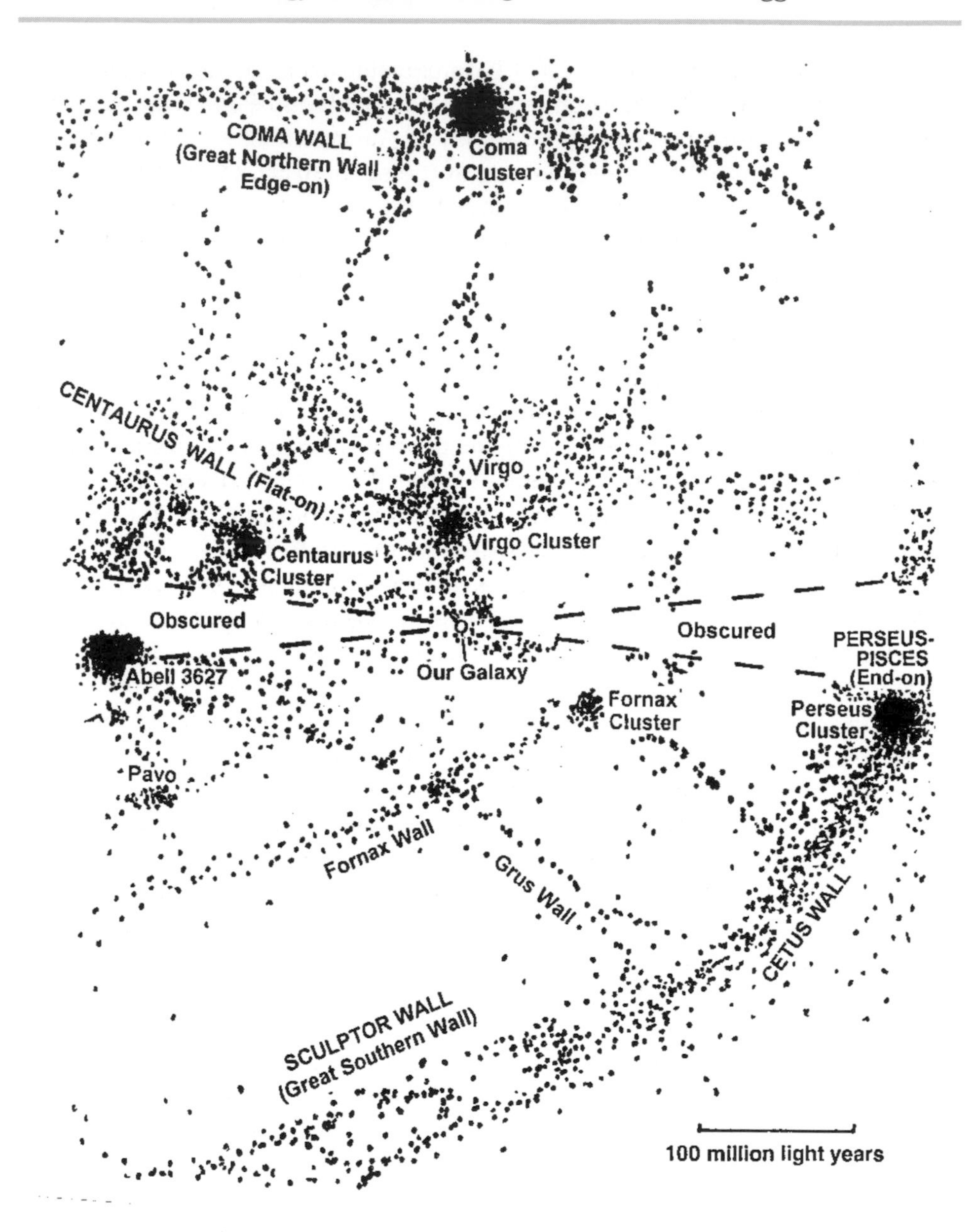

4.5 A schematic map of nearby large-scale structures. Although the layout is three-dimensional, the major structures can nevertheless be depicted on this "flattened" map.

The Local Supercluster is nevertheless not an independent entity in itself, but an appendage to a much larger structure, similarly flattened – a great wall! I refer to the structure as the *Centaurus Wall* (after the constellation it dominates); other investigators have called it the *HyperGalaxy*. It is seen as the most dominant feature in Map 5b (in the second color section). Figure 4.5 depicted it schematically flat on, with a rather irregular outline. The Centaurus Cluster, another well known concentration of galaxies, lies in this Centaurus Wall. However, the most dominant cluster in the structure is known, less romantically, as ACO 3627. This cluster was largely unrecognised until recently, because it is obscured by foreground Milky Way. It occurs just where the Centaurus Wall is seen to cross behind the plane of our Galaxy (see again the *Universe in 3D*). Like all large-scale structures, the Centaurus Wall has much irregularity in its form and shows numerous interconnections and even intersections with neighbouring structures.

Close by, some 50 million light years away, is a weaker, almost perpendicular structure I have called the *Fornax Wall* (others have called it the Southern Supercluster), because it contains the Fornax Cluster of galaxies. The Fornax cluster (again named for the foreground constellation) is not much more distant than the Virgo Cluster, but nowhere as rich.

About 100 million light years out, the *Hydra Wall* intersects the Centaurus Wall, also almost at right angles. This nearby structure is more of a "great ribbon" in form. It includes the Hydra Cluster of galaxies (also named for its foreground constellation).

Outside of these features, the texture of the Galaxy distribution is much lighter – until one encounters other massive features. At 200 million light years out, running almost perpendicular to the plane of our Centaurus Wall, but seen in the opposite direction, is a very dense filament – or great ribbon – the *Perseus-Pisces* Supercluster. Again the feature is named for the foreground constellations. Included in it is the Perseus Cluster. Perseus-Pisces has its interconnections. As Figure 4.5 suggests, it connects first to the *Cetus Wall*, and then to the *Sculptor Wall* in the South.

Finally, 350 million light years out, the last feature on the map is the *Coma Wall* or *Great Wall* – the first such structure to receive general recognition. It is centred on the rich Coma Cluster. Intermingled between all of these large-scale structures are numerous voids, one or two have already been mentioned – they are some thirty recognisable nearby voids, larger than a hundred million light years across.

So far we have described structures out to a few hundred million light years (those in Figure 4.5). While a number of more distant major structures have been recorded at greater distances, most are not yet mapped in detail. Only in certain directions have such detailed maps pushed further by means of dedicated redshift surveys. The *Las Campanas Redshift Survey*, together with the smaller European Southern Observatory *Slice Project*, were carried out in the late 1980s and early 1990s. Together they mapped some 30,000 galaxies to distances some ten times further than the Coma Wall. The surveys were

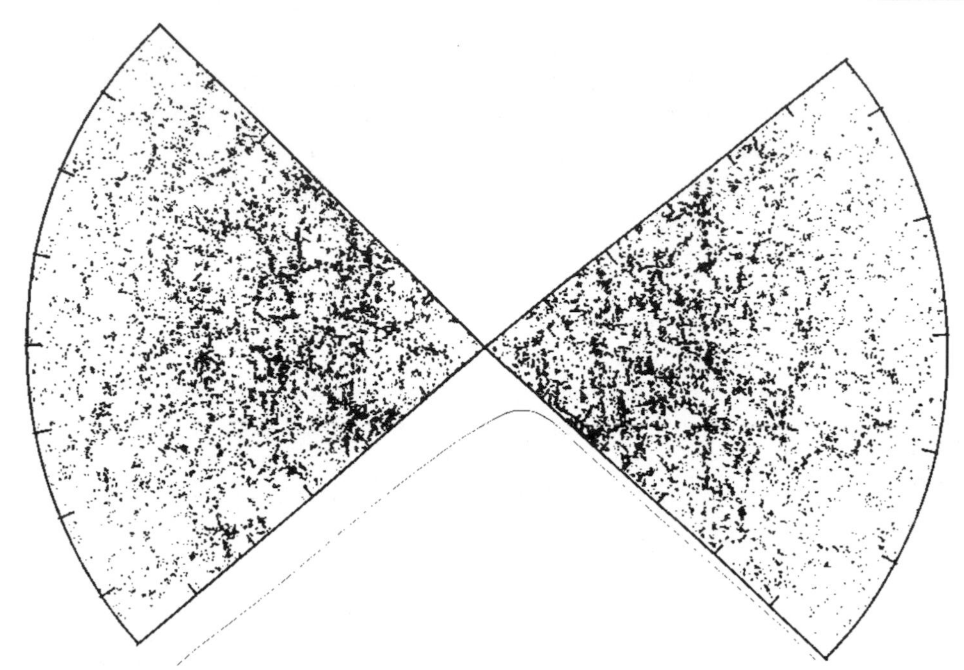

4.6 The Las Campanas Redshift Survey mapped the distribution of some 25,000 galaxies in thin slices stretching from our position (centre) out to two and a half billion light years.

carried out by taking adjacent narrow strips of the sky, so the outcome is more a two-dimensional cross section as is shown in Figure 4.6.

Perhaps the key question that these surveys sought to answer was whether we would encounter still larger, still more massive structures, or whether we would rather see repetition of the same sort of structures that we have described above. The Las Campanas survey suggests repetition. It has also been used to establish many fundamental statistical measures that scientists hope can then be used as discriminates in computer models that attempt to portray the evolution of the cosmos.

But scientists always want more data – even more than Las Campanas provided. So the hopes lie with two new surveys currently under way. The first is based at the Anglo-Australian Observatory and is known as the *2dF Survey* (for two-degree field). It revolves around a remarkable system that allows the telescope to gather the light of up to 400 galaxies simultaneously. The light is channeled via 400 optical fibres, the entrances to which are held on miniature magnetic buttons. A robot places the buttons on a metal plate, in appropriate positions for the particular field of galaxies to be observed (see Figure 4.7). The unit is then turned and placed where the telescope focuses its light. While the telescope observes the galaxies, slowly gathering enough light from each in an extended exposure, the robot busies itself setting up the next field.

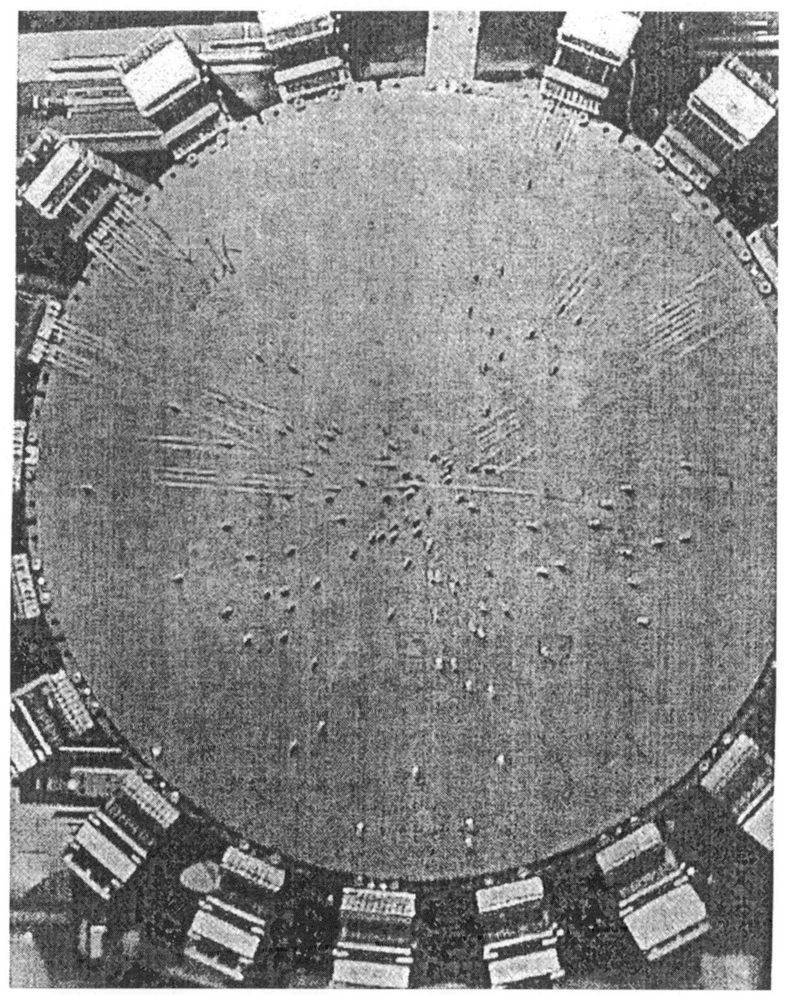

4.7 The Anglo-Australian Observatory 2dF system. 400 fibres have been accurately positioned to capture the light of 400 galaxies simultaneously (copyright AAO).

The second is the US–Japanese *Sloan Digital Survey*, which hopes to record the images of millions of galaxies, and obtain redshifts for perhaps one million, using a telescope dedicated for this purpose. Fibre optics are used to gather the light, but the robot simply drill holes in a metal plate, and the fibres are plugged in manually.

The 2dF and Sloan surveys are by far the largest efforts of their kind. Numerous other surveys have sought to penetrate deeper in small sample regions of the sky – usually small strips or, even simply, very small patches. The latter produce one-dimensional maps – much like bore-holes into the ground that seek to map underground geological formations. But they have the advantage of penetrating as deep into space as possible.

Ten years ago, such one-dimensional surveys revealed a very intriguing pattern. When aimed in directions perpendicular to the disk our Galaxy, they showed *periodic* structure – peak numbers of galaxies occurring every 600 million light years. Such periodicity, if rigidly true, would have thrown a spanner in the works of most cosmological theories. However, as more data has accumulated, the effect has faded, but it has not completely gone away. Most researchers now seem to accept it as a sort of characteristic size of large-scale structures.

Now that a number of very large telescopes have come into operation during the 1990s, it will be interesting to see if they can reveal anything about very distant large-scale structures back in cosmological time. Of course, the chances of being able to map like the Las Campanas survey are impossible – rather one has to make do with very tentative and fragmentary evidence, invariably probes in a single direction. Some intriguing "peaks" have shown up – suggesting the possibility of very massive structures.

Nevertheless one has to be careful. Additional data may make the peaks fade. Also, it is easier to successfully measure the redshifts of galaxies at particular intervals of redshift, so inevitably one is going to see some sort of peaks at those values.

Where did large-scale structures come from? Our image of the early Universe, on the inside shell of the Cosmic Egg, shows only vague irregularities (see back to Figure 2.5). These are somehow forerunners to the large-scale structures of today. However their scale is much larger, and we see them only as they were in the past, and not what they subsequently became. If, in the years ahead, we succeed in getting a more detailed picture of the irregularities in the Cosmic Microwave Background, then we may truly see the same sort of large scale structures of today, but in embryonic form.

One way of investigating how the early irregularities evolved to the large-scale structures of today is to use computer simulations. This, in itself, has given rise to a whole research industry. If you have seen the Imax film *Cosmic Voyage,* then you saw one such simulation. That these simulations produce something like the large-scale structures of today, show that we are probably on the right track (but more will be said in Chapter 7).

Still the ultimate question in large-scale structures has to be *where did their pattern come from in the first place*? What wove the cosmic tapestry? Where did the texture itself originate? Though science thought it had the answer, recent developments – and perhaps my own intuition – suggest that we cannot be so sure. But we can speculate and indeed will be doing so a little later on, again in Chapter 7.

Chapter Five

Long Ago and Far Away

Long ago I saw a movie based on H.G. Wells' *The Time Machine*. It is memorable because the concept of being able to travel back and forth in time is a very appealing one. Imagine whisking back in time to watch the pyramids being built, or even the Wright bothers making that first flight. Would the future prove reassuring or scary? Would we be able to tamper with events to change the course of history?

That only happens in the movies. Reality is that *we cannot look into the future*. Of course, it is surprising how many people are taken in by fortunetellers and astrologers. As an astronomer, I have on occasions found myself in contest with astrologers. It would, of course, be delightful if the true purpose of the planets wheeling around the Sun was to tell us whether or not it was a good day for dealings with friends. But I am far from convinced. Even the "sign of the Zodiac" I am allocated in the weekend newspapers is incorrect astronomically. If an astrologer claims that Jupiter is in "Sagittarius", you can go outside at night and find it in the constellation of Scorpius! Two and a half thousand years back, the astrological signs matched the starry constellations. Thanks to a slow wobbling motion of the Earth, they do so no longer – or rather will only do so in 23,500 years time. For most of time, astrologers somehow happily work without stars!

We may not be able to gaze into the future, but we can very easily look into the past. For that purpose, ever more powerful time machines were built during the course of the 20th century. Machines that can take us to different times. But they are not the sort that H.G. Wells had in mind, and they can only operate back into the past, not into the future. Furthermore, you cannot use them to go back and change anything that has already happened, but you can at least see how things *used to be* in the past.

Of course, it is a little misleading to call them "time machines" instead of "telescopes". But large telescopes are in fact time machines. We have already seen (particularly in Chapter 2) how by looking deeper into space, *we look back into time*. Consequently we see the content of our Galaxy, not as it is today, but as it was thousands of years ago. We see companion galaxies, not as they are today, but as they were millions of years ago. We see clusters of galaxies, not as they are today, but as they were billions of years ago. Eventually we see the very early Universe as it was fifteen billion years ago – the opaque shell of the Cosmic Egg. Large telescopes let us see back many billions of years, deep into cosmological time. They open up the great history book of the Universe.

What a wonderful opportunity! What a monumental advance in our exploration of the observable Universe. It has almost all come about in the 20th century; most of it has happened in the last decade.

For most of the existence of the human race, we had no telescopes – only the naked eye. How did this affect our perception of the Universe about us? Let's see the score. Of the other seven major planets in the Solar System, five are visible to the eye. Of the countless minor planets and bits and pieces in the Solar System, only those that collide with the Earth are seen as shooting stars, and only the icy ones that go too close to the Sun create visible comets. Of the billions of stars in our Galaxy, only several thousand can be seen with the eye. Finally, (as pointed out in the previous chapter) of the billions of galaxies in the Universe, only *three* outside our own are visible to the naked eye. The naked-eye view of the night sky is very beautiful, particularly away from city lights and when the bright Moon is absent, but it is a very shallow view of the Universe.

The reason why the eye does not see more planets, more stars and more galaxies, is that the rest are *too faint*. To see fainter things, you need to gather more *photons* of light – and that calls for bigger eyes! Telescopes are bigger eyes. Just compare the difference. When your eye is fully dark-adapted (each pupil open as wide as possible) its aperture is about 6 mm across. Even a very modest amateur telescope may have an aperture 60 mm across. You do not need to be much of a mathematician to realise that a small telescope collects a *hundred times* more photons of light. It is no different to putting an eggcup and a pie dish out in the rain – the pie dish will collect many more raindrops than the eggcup.

Telescopes are like pie dishes – they let us see fainter objects because they gather more light. The bigger their aperture, the fainter the object they see, and the deeper into space they penetrate (see Figure 5.1). So the general progress this past century has been to try to build ever larger telescopes. At the end of the 19th century, the largest telescopes had apertures of about a metre. At the end of the 20th century, the largest telescopes had apertures of

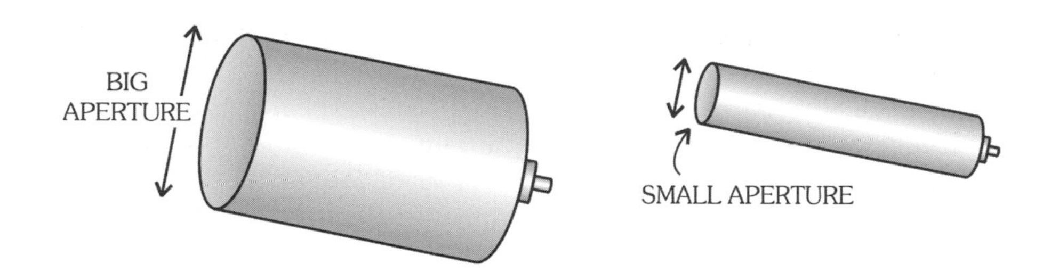

5.1 The size of a telescope is gauged by its aperture. The larger the aperture, the fainter the objects it can detect.

about ten metres. Put another way, the largest telescope today can collect about three million times more light than can the unaided human eye!

A pie dish is not a bad analogy for a telescope, because a modern large telescope uses a large *concave mirror* to collect the incoming light and direct it to a focus. Up to a hundred years ago, the largest telescopes were still *refractors* – where a *lens* collects light – a sort of scaled-up version of the telescope that Nelson raised to his blind eye. But mirrors have proved to be better telescope *objectives*, because they only require a single surface to be figured to the correct shape. Also, mechanically, they can be supported from behind. All modern telescopes use glass mirrors with a shiny coating of aluminium or silver. The light reflects off the concave surface, without passing through the glass. Years ago, mirrors were made from polished metal, but glass has since proved a more suitable material as it is less susceptible to thermal expansion.

Of course, it is being somewhat derogatory to label a telescope mirror a pie dish. For it to function properly, its surface must usually be shaped correctly to an accuracy of a ten thousandth of a millimetre – and that is the challenge of the optical engineering of a telescope.

So the ability to gather more light is the first good reason for building telescopes. The other good reason is more obvious – to magnify. Telescopes increase the detail that we can see – they make things look larger to our eyes.

But whether you are looking in the sky for birds or stars, it is the *angular size* that is relevant. It happens, for example, that the Sun and the Moon appear to be about the same angular size in the sky. In reality, the Sun is much bigger than the Moon. However, while it is about 400 times the Moon's diameter, it is also 400 times more distant. So it appears the same angular size – about half a degree across. Telescopes let you *increase* that angular size. If your pair of binoculars (two folded telescopes side by side) says 7×50, then it will magnify the angular size by 7 times (and the diameter of the front lens, its aperture, will be 50 mm). You can check its magnification by holding only one side to your eye and looking at a brick wall with both eyes open. You can then easily count how many "naked-eye bricks" fit into one "through-the-binoculars brick". Aim those binoculars at the Moon and its angular size will be magnified to three and a half degrees. But do not try it with the Sun!

From the horizon to the point overhead is, of course, 90 degrees – so the Moon's angular size, only half a degree to the unaided eye, is only 1/180 of that. For useful reference, note that, when held at arm's length, the width of your finger is about 2 degrees, and your outstretched hand (thumb and little finger as far apart as possible) about 20 degrees. By the way, the human eye is notoriously inaccurate in judging angular sizes. Everybody swears the Moon is much bigger when they see it rising on the horizon, than when it is higher in the sky – but that's an optical illusion!

Astronomers also work with very small angles. Like the *hour*, a *degre*e can be subdivided into sixty *arc minutes*, and an arc minute into sixty *arc seconds*. The smallest detail the eye can *resolve* is about two arc minutes. That is because there are lots of microscopic-sized detectors (known as cones and rods) in the retina of the eye. They work much like the pixels on a computer screen – each detector operates as one pixel. They are packed the closest in the eye's central spot, where the eye has the greatest resolution. The eye could, of course, do better if the image on the retina was enlarged. That is where a telescope comes in.

For example, our nearest neighbouring solar system (as encountered in Chapters 1 and 2) is *Alpha Centauri*. It has two bright suns that, seen from our distance, are about a quarter arc minute apart in the sky. Look at that star with the naked eye, and it appears as though it were a single body. Look at it through a small telescope, and you can see it is double.

In similar fashion, you can see craters on the Moon through a telescope, but you cannot resolve them with the naked eye. Or you could choose to read your Sunday newspaper from 20 metres away with a telescope.

So telescopes are also made to magnify. In fact, the magnification can be varied on most telescopes by changing eyepieces. Low magnifications are fine for looking at starfields – since, even under high magnification, stars are still pinpoints of light. High magnifications are good for looking at the Moon and planets, but only if the atmosphere allows it.

The highest practical magnifications to use – no matter how big the telescope – are never more than several hundred times, for the atmosphere of the Earth makes it senseless to try to magnify further. You have probably looked on a hot day and seen distant objects shimmering. Well, even when the atmosphere is much calmer, it still has a blurring effect on everything we see through it. Stars ought to be seen as pinpoints of light; but look through a telescope, even with excellent optics, and you will see them as small blurry points, with angular sizes usually an arc second or two.

To illustrate the point, this segment of the text is being written while the author is sitting observing at a telescope at the South African Astronomical Observatory, situated in the semi-arid Karoo in the Western Cape, near a small town called Sutherland. "Observing" is a slight over-statement; the telescope is on a type of auto-pilot, and the author, sitting in the control room, can get on with other jobs, such as writing this book. At the moment the star images appear about one arc second in size. A few nights ago, they were even better – smaller than an arc second. But during the night that followed, just prior to a short dose of inclement weather, they blew up to two to three arc seconds. Astronomers refer to this as the *seeing* – good seeing when the star images are small and sharp, poor seeing when they are big and blurry.

Since large professional telescopes are expensive pieces of equipment, it pays to site them where the observing conditions are best, with as many clear nights as possible and with as good seeing as possible. High humidity is often the cause of bad seeing, so astronomers and their telescopes migrate to drier climates (which explains the total absence of trees outside). But extreme deserts are not suitable, as often too much dust is airborne – and dust is bad for telescope optics and observations. So usually, the answer is arid, but not desert. That puts most of the world's large telescopes around latitudes thirty degrees north or south of the Equator, and toward the western edges of the continents.

Or, you can find a mountain reaching high above a cold ocean. Both Hawaii and the Canary Islands – dormant volcanoes pushing up from the ocean floor – have proved winners. The cold ocean helps stabilise the atmosphere. High altitude is another major prerequisite – it decreases the amount of air between the telescopes and outer space.

It is also desirable for the telescopes to work with as dark a night sky as possible. This is also important for seeing very faint things. The Moon is the problem. It may be a beautiful romantic orb to lovers, but it is a great nuisance for astronomers! Just as sunlight is scattered in the daytime sky, so moonlight is scattered in the night-time sky. Happily, the Moon is not always above the horizon when astronomers are trying to work. But work at observatories is according to the phase of the Moon. Right at the moment, here where I am observing, it is prime time, as the Moon is very close to New Moon. The work I am doing involves going as faint as practical – by letting the telescope expose on the same small region of sky for an hour. It can only be done at this time of the Moon's cycle.

Light pollution is another matter. Only fifty years back, most observatories were located, for convenience, inside cities. However, great "breakthroughs" in street lighting were made in the 1960s, with the development of more efficient lamp bulbs. The amount of outdoor light generated within cities has escalated ever since. And cities have grown with the doubling of the world's human population. It would be fine if all the light only went downward, or only lit up what it was intended to do. However, a significant amount escapes upward into the night sky and is scattered. You need only try looking at the night sky from within a city, to realise that you cannot see the fainter stars.

Efforts are under way to improve things, thanks partly to the activities of the *International Dark Sky Association.* This may seem surprising, as most other pleas by astronomers usually fall on deaf ears. Fortunately, wasted light going upward is wasted electricity and wasted money. Cities, seeking to save, realise the sense in using light fixtures that only illuminate what is needed and do not shine into space. Already some cities have significantly reduced light pollution. The leader (not surprisingly also the headquarters for the International Dark Sky Association) is Tucson, Arizona – since it has the major U.S. National Observatory at Kitt Peak only about 60 km from its downtown. The choice of siting the observatory there was made around 1960 – when Tucson was only a small town and streetlights had still to grow up.

At about the same time as that observatory was being founded, the astronomical world was rocked by the discovery of *quasars*. The word is an acronym for "quasi-stellar radio sources". Of the billions of stars visible to telescopes, these objects were initially singled out because of their radio emission. But they proved not to be stars – instead they were *incredibly distant galaxies*. Normally such distant galaxies would have appeared as very, very faint blurs, typically at the limit of a telescope's sensitivity. Yet these were bright and very clearly visible.

The light of a quasar comes from the very centre of its host galaxy. It is not normal starlight; nor could stars generate so much light in so small a volume of space. We believe that the quasar phenomenon is a manifestation of matter being violently accelerated as it is sucked in toward a massive black hole (about which more follows later in Chapter 8). A small amount of matter gains enormous energy, a part of which powers the quasar. We now recognise the same process occurring on more modest scales, even in nearby galaxies. A class known as *Seyfert galaxies* (after their discovery by Carl Seyfert in 1943) are generally milder forms of quasars. There is even good evidence that our own Galaxy harbours a massive black hole at its centre, but, like a dormant volcano, it is inactive at present. In any case, our Earth is situated far from the centre of the Galaxy, so we are keeping a safe distance.

In terms of sustained luminosity, the most extreme quasars are the brightest objects in the Universe. Or, rather, they used to be. We can comfortably see quasars many billions of light years out, and therefore many billions of years back in time. They seem to be at their densest about two thirds of the way out to the shell of the Cosmic Egg. That was when the Universe was only about 5 billion years old, and galaxies were packed into much smaller volumes of space. Many investigators feel that almost all giant galaxies, ours included, have been through periods of quasar activity. Most likely, close brushes with other galaxies, which stirred up interstellar material, and set it falling toward the central black hole, initiated these.

Quasar activity still occurs today, but on a very reduced scale. The encounters that excite it are now much rarer, as the separations between galaxies are much greater. But it also declines as one looks toward earlier epochs. Perhaps the black holes then still had to gain much more mass to be effective, or perhaps there were not so many galaxies. Nevertheless, we can see quasars out to some 85 percent of the distance to the Cosmic Egg – as illustrated in Figure 5.2.

The data shown in the diagram comes from a survey being carried out with the Anglo-Australian Telescope, using the same equipment (shown in Figure 4.7) used to get redshifts of galaxies. With quasars, it looks very much deeper into space, and is picking up very large numbers and mapping their distribution.

Up to only a few years ago, quasars were all we could see that far out. The normal galaxies out there – if there were any – were just too faint for our telescopes. However, recent developments have changed all that. We now live in exciting times, as a new generation of very large aperture telescopes has come into operation. Moreover, we have the Hubble Space Telescope.

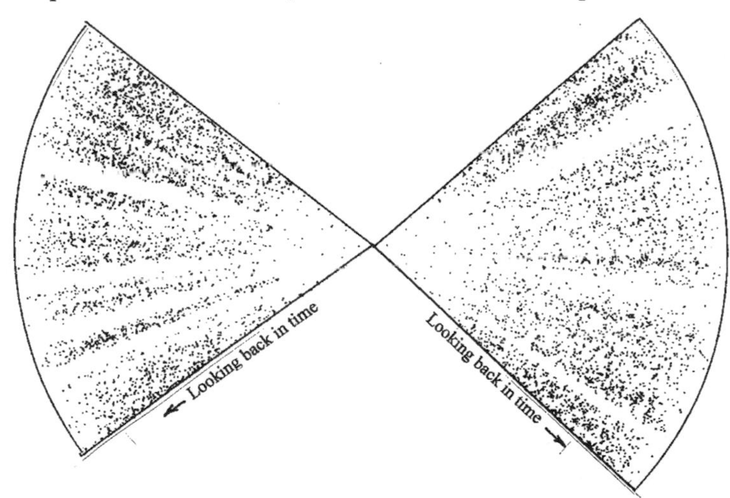

5.2 The distribution of quasars in two thin slices centred on our Galaxy, from the "2dF" survey carried out with the Anglo-Australian Telescope. The radial striations are not real, but are due to incompleteness in the directions surveyed. The radial variation is real – the number of quasars decline toward the present (centre of the diagram), and toward the very early Universe (edges of the diagram) (copyright AAO).

Launched in 1990 from the Space Shuttle, the Space Telescope (named in honour of Edwin Hubble) initially received much adverse publicity – and the anguish of its intended users – when it was found that a mistake had been made in the final figuring of its mirror. However, the first of the Shuttle servicing missions, in late 1993, was able to fit the telescope with corrective optics – like a pair of spectacles – to overcome the problem. Since then, it has become the "discovery machine" it was designed to be.

The size of the Space Telescope is very modest, only 2.4 metres in aperture, since it had to fit into the cargo bay of the Shuttle. It is also an old-fashioned telescope, designed in the 1970s. Unlike modern telescopes, its main mirror is rigid and its exact shape cannot be adjusted, otherwise it would have been easy to have corrected the mistake in the mirror's optics.

Putting anything into space is extremely costly – often about a hundred to a thousand times more expensive than an equivalent facility on the surface of the Earth. So it has been with the Space Telescope, which has probably already cost more than everything else in astronomy. It is said that for the cost of a single instrument that mounts on that telescope, you could replace many of the major observatories on the surface of the Earth!

But the great advantage of the Space Telescope is its resolution – down to a tenth of an arc second! This is possible because it does not have to look through the Earth's atmosphere. As described above, that atmosphere normally blurs the view, turning each star image into a tiny fuzzy ball. Above the atmosphere, the only limits are the quality of the optics and the precision with which one can hold the telescope still. Hence the need to correct the faulty optics, while the "fine guidance sensors" on the Space Telescope enable the vehicle to hold itself incredibly steady.

But a challenge is coming from the ground. Very rapid movements in the position of a stellar image cause much of the atmospheric blurring. Modern technology allows these movements to be monitored, and *active optics* – where a computer can rapidly deform or tilt a mirror to compensate for the effects of the atmosphere – is catching up. Images with resolutions close on 0.2 arc second have been achieved.

The high resolution, but small aperture, of the Space Telescope has been complemented by the building of the new generation of very large ground-based telescopes, with apertures of some eight to ten metres.

The Keck telescopes (named for the main benefactor), are twins sited on the summit of Mauna Kea on the Big Island of Hawaii. They were the first operational telescopes of this new generation. Their design was considered revolutionary because their primary mirrors – the concave pie dishes – are each made from 36 separate segments. The idea had been tried before – it is obviously easier to make small mirrors – but never successfully. The difficulty

5.3 The twin domes of the Keck telescopes (Photo: W. M. Keck Observatory).

lies in positioning the segments to the accuracy involved, complicated by the fact that the supporting structure unavoidably deforms, slightly but significantly, as the telescope moves. However, it is possible to have each mirror positioned under computerised control. An array of sensors between the mirror and its neighbours feels the relative position, and feeds the information into the controlling computer. As the telescope moves, or tracks objects in the sky, so the system is constantly adjusting the supports to the segments to keep them operating with the precision of a single giant mirror. Once the first of the Keck telescopes was operational, and the technique proved successful, a second telescope was started. As seen in Figure 5.3, the two telescopes are sited close together, the plan being to link them on occasions, so as to operate as a single instrument, with remarkable resolving power.

Segmented mirrors are a great cost saving, and the revolutionary Hobby-Eberly telescope in Texas – built at a small fraction of normal cost – is another ten-metre class telescope (see Figure 5.4). Its primary mirror consists of 91 segments. Having spherical curvature, the mirrors are identical. All are computer controlled and aligned to a common focus. Unlike all other large optical telescopes, this telescope does not move or track while observing. Instead, a "tracker" unit, moving about at the top end of the telescope does the job. Again a twin is being built, but in the Southern Hemisphere so as to have access to the southern skies, at the South African Astronomical Observatory.

5.4 The Hobby Eberly Telescope (Photo – McDonald Observatory).

An alternative approach is to have a single large thin meniscus mirror. The mirror is so thin that it would break under its own weight, were it not always carefully supported. Unlike the old-fashioned approach of having a heavy rigid thick mirror, the meniscus is flexible enough that an active computerised support system can hold it to a precise figure. The largest meniscus mirrors that can be comfortably made at present are eight metres in diameter. Consequently, we are seeing a number of such telescopes coming into service.

The Very Large Telescope of the European Southern Observatory (Figure 5.5) consists of four such units, sited on a mountaintop at Cerro Paranal in the

5.5 One of the four units that make up the European Southern Observatory's Very Large Telescope (Copyright ESO).

dry northern desert of Chile. Like the Keck telescopes, the units can operate independently, but the plan is also to have them work on occasions in unison, linked as a giant "interferometer" - one enormous telescope.

Gemini North (on Hawaii's Mauna Kea) and Gemini South (on Chile's Cerro

5.6 The interior of the Gemini North Telescope, with its viewing and ventilation shutters open. (Photograph by Neelon Crawford/Polar Fine Arts 1999. Copyright Gemini Observatory)

Pachon) are similar 8-metre class telescopes – a collaboration between the United States, Britain, Canada, Chile, Australia, Argentina and Brazil (see Figure 5.6). Subaru, the prime research facility for Japan is another 8-metre telescope situated on Hawaii (see Figure 5.7).

Never before have we possessed such powerful instrumentation, never before have we seen so clearly deep into space – and deep back into time. We can look upon the Universe as it used to be. Now, almost every month, claims of a new most distantly observed galaxy, or most distantly observed quasar are made.

The Hubble Space Telescope led the way in 1995. Robert Wilson, then the director of its Space Telescope Science Institute, used his discretion, and his share of observing time, to have the telescope observe the same small patch of sky for over 10 days. In that time, hundreds of individual exposures were made, through various different colour filters, and afterwards all summed together to make the deepest photograph ever taken.

A black and white version of that photograph is shown in Figure 5.8 (but a full colour version is found in the first colour section). The full photograph covers only a tiny part of the sky (less than 1 percent of the size of the Full Moon). Since it is aimed almost directly out of the Galaxy, it shows hardly any

5.7 The Subaru Telescope (Copyright 1999, Astronomical Observatory of Japan)

stars (which would lie in our Galaxy) at all. But it reveals some 3,000 other galaxies. Even the other nearest of these galaxies – those with the largest angular sizes – would have been considered very distant by older standards. But the relatively small number of nearby galaxies are greatly outnumbered by the fainter galaxies. The average distance of the galaxies in this picture is an astounding *several billion light years*. Some, of course, are much closer than that, many are even more distant, perhaps out to twelve billion light years.

And so we see our Universe of galaxies, not as it is, but as it was. Several

5.8 A portion of the Hubble Space Telescope Deep Field.

billion years means that we look back roughly to half its present age. At first glance, it does not look all that different to the present. Both spiral and elliptical galaxies have been discerned. But there are differences. For one, there are many more faint galaxies than there ought to be. This has told us that the era of star formation in galaxies reached a peak in the past, and is now on the decline. But where have all those faint galaxies gone? The answer is either they have fizzled out or they have merged with others. We have enough evidence of galaxy mergers to know it happens on a significant scale. Back in the past, when galaxies were much more crowded together, it must have been far more frequent. Alternatively, many of the galaxies could still be around today, but their starlight has faded to insignificance. Indeed we do suspect a hidden population of "ghost" galaxies – galaxies long over the hill as far as star formation is concerned.

A general characteristic of the galaxies in the Hubble Deep Field is their more irregular nature, compared to the smoother more symmetrical forms of today. Again, this suggests disturbances due to the close proximity of neighbouring galaxies. Some galaxies, however, look very much the same then as they might today. It suggests that we are still not looking back quite far enough. Both this evidence, and the presence of quasars, show us that galaxies must have formed relatively early in the Universe – well out toward the shell of the *Cosmic Egg*.

While the Space Telescope reveals the greatest detail, it lacks the light gathering power to analyse the physical nature of the distant galaxies. This is where the Keck telescopes, and the more recently commissioned large-aperture instruments, come in. They may not see as much detail, but they can break the light into a spectrum to tell us whether we are seeing a normal population of stars, low density gas, or quasar activity. The most distant known galaxies have generally been examined by the Space Telescope and by the new large-aperture telescopes.

The Hubble Space Telescope has completed a second deep field (as mentioned earlier on page 34), in the same fashion as the first, but this time looking in roughly the opposite direction. It looks much like the first, and it is reassuring that the Universe looks so similar, even in opposite directions (again see the first colour section in this book).

The similarity still holds, even at other wavelengths. We already know that the same applies to the microwaves that emanate from the inside shell of the Cosmic Egg; one side looks much like the other. Still other wavelengths have been explored. The numbers of galaxies that emit significant radio waves look the same in opposite directions, but they do vary considerably as one looks back through time. Again, one sees such activity peak in the past. We are also beginning to explore the distant cosmos at other exotic wavelengths, even X-ray and gamma ray.

The Universe continues to surprise us. In very recent years, enigmatic *gamma-ray bursters* have now been traced to extremely distant galaxies. Within those galaxies some traumatic events occurred that gave rise to a brief burst of radiation, which in energy far outshone even the quasars.

The story is by no means over. A large aperture telescope in space – the Next-Generation Space Telescope – is now the dream. On the ground, the success of the segmented mirrors has led to speculation of giant optical telescopes with apertures approaching a hundred metres! The quest to build ever-larger telescopes – ever better time machines – continues apace.

Such instruments will reveal still fainter objects, still closer to the ultimate horizon that is the shell of the Cosmic Egg, as they explore the Universe of long ago and far away!

Chapter Six

A Universe in Motion

A well known children's story tells of Chicken Licken who rushed around spreading the news that "the sky was falling". No wonder she was confused; it is not the sky that is falling, it is us!

In the 1970's, measurements of the "picture" on its inside shell of the Cosmic Egg showed we were falling. We are falling on a cosmic scale, and thanks to the expansion of the universe, we are not about to hit the ground, or anything else.

Our falling is just our ultimate motion within a Universe constantly in motion. If you thought you were sitting quite stationary reading this book, it's an illusion. *You are constantly on the move.*

For a start, the Earth is spinning. The cycle of day and night is, of course, due to the rotation of the Earth. That rotation carries you once around a circle in almost 24 hours. The further you are from the North or South Poles, the faster you move. If you were on the equator, you would be travelling at a speed of 1,670 kilometres per hour, or 460 metres per second – slightly faster than the speed of a bullet! Further from the equator, and you go a little slower. This book is being written at a Latitude 34 degrees South of the equator, where the speed works out at 1,380 kilometres per hour, or 380 metres per second.

It is like being on a fairground merry-go-round, but much more difficult to step off. A pity – otherwise, you might for instance "step off" from Lisbon, wait as the world turns for four and a half hours, then step back on at Washington D.C. After a brief look, step off again for two hours and back on again in Denver. Off for a further ten hours, and you could have a brief look at Outer Mongolia. Wait another seven and a half hours and you could be back where you started in Lisbon. A great way to see the world without going anywhere!

As with the merry-go-round, it is very easy to imagine yourself as stationary and the rest of the world as turning around! It may seem that the horse that you ride is stationary, and the rest of the fairground, houses, trees and all are turning, turning. Most people consider themselves stationary, and do not think of themselves as constantly going around and around.

We talk of "*sunrise*" when the Sun does no such thing! The Sun is to the Earth like an elephant is to a mouse. It is as though the mouse turned around, yet claimed it was the elephant that moved!

The Sun doesn't rise. It is the horizon that sinks! Of course, "horizon-sink" doesn't sound anywhere as impressive, or as poetic, as "sunrise". At the end of the day, it is the other horizon that rises, not the Sun that sets. Again "horizon-rise" doesn't have as good a ring to it as "sunset". But it is not just at sunrise and sunset that the eastern horizon is sinking and the western horizon rises. It happens all the time. If you look toward the eastern horizon, you ought to be able to realise that it is dropping and the ground on which you stand is tilting over toward it. Can you feel yourself tipping over?

Obviously not. Human senses do not feel the motion. Primitive man and ancient societies – Egyptians, Babylonians, Romans, Mayan – all adopted the idea that the Earth was *still*, and that the Sun and other heavenly bodies *moved* across the sky. While some forward thinking ancient Greeks contemplated a turning Earth, nobody could *feel* the Earth turning. Since there was no strong wind constantly from the east to blow the trees down, nobody could sense we were turning. The Earth seemed stationary.

The concept went even further. *Heavens above!* But never *heavens below*. That was usually reserved for the other place. Yet, any modern person should be aware of the Earth being spherical. Consequently, the *heavens above* in Europe are the *heavens below* down under in Australia. And vice-versa. Also, since the Earth is turning, the *heavens above* generally become the *heavens below* after some twelve hours.

It is not only the Sun that appears to rise and set. The Moon does it – at all hours of the day and night, depending on its phase. Stars do it! At night, stars are constantly rising all along the eastern half of the horizon, and setting all along the western half. And there are as many stars below the horizon as there are above. After all, if you could imagine yourself floating in space, you ought to be able to see stars in every direction, *whether up or down*. On Earth, you look *up* into space, but not down. With a flat horizon, you get to see half the sky. But, as the Earth rotates, you see some stars apparently rising, some apparently setting.

It is not to say that the half of the sky below the horizon becomes the half of the sky above the horizon after the Earth has rotated through 180 degrees. Unless you live at the equator. If, on the other hand, you lived at the North or

South Pole, then the half that is below the horizon stays below the horizon, and the half that is above the horizon stays above. In short, stars do not rise or set at the poles – except for one bright star, the Sun, and that is a complication we will not get into for the moment.

Most of us live neither on the equator nor at the poles, and we get an intermediate situation. Most stars apparently rise in the east and set in the west, but there is a portion of the sky that never sets – the stars in it are permanently above the horizon. There is also a portion of the sky that never rises – with stars permanently below the horizon.

Figure 6.1 illustrates the situation. In the figure the starry sky around the Earth has been represented by a sphere – known as the *Celestial Sphere.* Though based on the ancient concept – promoted by Greek civilisation – that the stars were literally fixed upon a sphere, it is retained nowadays as a convenient, albeit fictitious concept for representing the sky; one sees in it a forerunner of the Cosmic Egg.

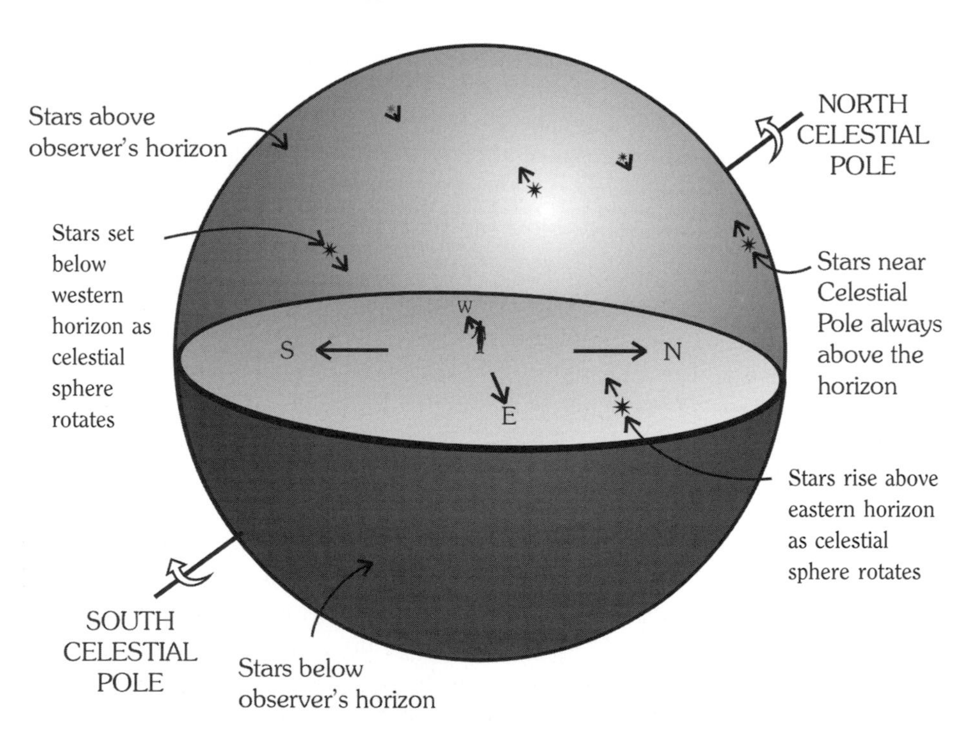

6.1 The Earth's rotation is reflected by the apparent movement of the starry sky. The observer's flat horizon can be extended to meet an imaginary celestial sphere, upon which are "fixed" the stars. The diagram depicts the situation for middle latitude in the northern hemisphere; in the southern hemisphere, the south Celestial Pole would instead be elevated above the southern horizon.

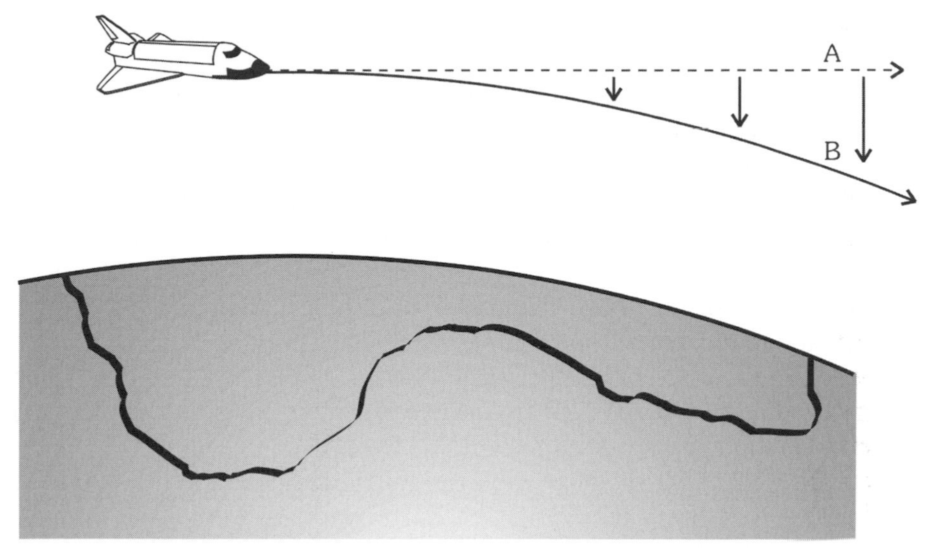

6.2 Were it not for gravity, the Space Shuttle would travel in a straight line. However, the Earth's gravitational pull causes it to fall, so it follows a curved path. Provided it can move sideways fast enough, it never hits the ground.

Where the ancients supposed there was a celestial sphere, there is today a swarm of thousands of artificial satellites and spacecraft. These satellites are always on the move. Otherwise they would fall to the ground. The principle of orbiting a spacecraft is to set it moving "sideways" fast enough, so that it falls but never reaches the ground (see Figure 6.2). Its trajectory simply follows round with the curvature of the Earth's surface. Given that there is no drag in the vacuum of space, the spacecraft does not lose energy and maintains the same average distance from the centre of the Earth. It takes an awful lot of rocket fuel to get a satellite up into orbit, but once there, it coasts without effort.

Most of the spacecraft that immediately come to mind – the Space Shuttle, Mir, the International Space Station – are in *low* Earth orbit. They travel around the Earth in circular orbits, only a few hundred kilometres above the ground – literally skimming the top of the atmosphere. To do this, they must be accelerated sideways to a speed of 8 kilometres per second (5 miles per second). At that speed, it takes only 40 minutes to get from Europe to Australia! Or around the Earth in just an hour and a half.

Satellites that skim the top of the atmosphere pay a price. Once every eleven years (Year 2000 for instance), the Sun reaches a peak of activity. A consequence of this is that the upper atmosphere gets heated and swells upwards – putting a slight drag on the satellites. Once a satellite is slowed, it descends and reenters the atmosphere.

To throw a stone – or spacecraft – *higher* takes *more energy*. To orbit higher, more rocket power (which costs more money) is needed. The Space Shuttle itself does not have the power to go higher. Even if you do get something up higher, its eventual speed is *slower* and, because the orbit is bigger, the time to go around once is much longer. This enables us to put some satellites high enough – nearly 40,000 kilometres – that they take 24 hours to orbit, the same as the Earth takes to spin. Those satellites therefore always look down on the same side of Earth.

In a still much higher orbit – and travelling at only a kilometre per second – is the Moon, the Earth's only natural satellite. At an average distance of almost 400,000 kilometres, the orbital period is over 27 days. The Moon also rotates in the same time that it revolves around the Earth, so we get to see the same face of the Moon always turned toward the Earth.

In the same way that satellites orbit the Earth, so the Earth orbits the Sun. Seen looking down from the North, the Earth revolves around the Sun in an anti-clockwise direction – it also rotates anti-clockwise. The orbital period is now a year. Again it's a motion we cannot feel. As we travel around the Sun, we are moving at 30 kilometres per second – more than a hundred times faster than a Jumbo jet flies.

The knowledge that the Earth goes around the Sun is relatively recent. In spite of early speculation by the Greeks, it was only proved so in the last few hundred years. Most older civilisations considered the Earth stationary, and thought the Sun moved. The Sun was often likened to some sort of powerful god, and its "movement" against the constellations gave rise to the "Sun signs" of Astrology – the twelve "signs of the Zodiac".

The cycle of the seasons – the origin of the *calendar year* – has been known since antiquity. We now understand it to be a consequence of the 23.5 degree tilt of the Earth's axis, as our planet travels around the Sun. Though it is common experience, and supposedly widely taught at school level, relatively few people can explain what causes the *seasons*. The simple question "Why is it hotter in summer than in winter?" invariably brings erroneous answers such as "because the Earth is closer to the Sun". That may be partially correct in the Southern Hemisphere, since the Earth is slightly closer to the Sun in January, but it hardly works for the North. The correct answer is to do with the *angle* at which sunlight strikes the ground. When sunlight strikes the ground *obliquely*, it gets spread much thinner (like using only a little butter to cover a lot of toast), and its heating effect is diminished. Put another way, the *higher* the Sun is in the sky, the greater its heating effect. The Sun is generally low in winter, and high in summer – so summers are hotter. Also, regardless of the season, the Sun is low early morning and late afternoon – and its heating effect is decreased then as well.

The Earth is not the only body to orbit the Sun. There are seven other big planets and a multitude of smaller bodies, all constantly in motion. And the same rules that applied to satellites going around the Earth also apply to planets going around the Sun: The further the planet is from the Sun, the slower it moves in its orbit, and the longer it takes to go around. For instance Mars, next out from here, takes almost 2 years – and one imagines hypothetical Martian tax payers only making their returns once in 2 years! Birthday presents are obviously rare on Jupiter, where you would have to wait out patiently one "Jupiter year" of almost twelve "Earth years". At the extremes, little planet Pluto and friends move at only 5 kilometres per second and take 248 years to go round once, whereas Mercury, closest to the Sun, hurtles along at 50 kilometres per second and has four "years" to one of ours.

Viewed from a moving Earth, these other planets appear to follow complex motions in the night sky. Known as "planetes" – the Greek for "wanderers" – they seem to trace paths against the starry background that often involve loops. Understanding how these are produced has been the obsession of astronomers of old. Only a few hundred years ago, with the insistent belief that the Earth be the centre of the universe, a great system of revolving concentric crystalline spheres was hypothesised, to explain the motions of the planets. Eventually, the motion was correctly explained by the Sun being in the centre of the family of planets, and the spheres discarded – thanks to Nicolas Copernicus, Tycho Brahe, Johann Kepler, Galileo Galilei and, of course, Isaac Newton.

Solar-like systems exist within our Solar System. It was Galileo who first trained a telescope on Jupiter and discovered it had four large moons in motion about it. The innermost of the four goes around Jupiter in almost two days (Earth days), and the outermost in nearly seventeen. Jupiter also possesses numerous tiny moons – some captured asteroids – while swarms of fine particles orbit the planet closer in, forming a ring around the planet. It is like a miniature solar system in itself. And the same goes for Saturn, Uranus and Neptune. Each has a number of moons – large and small – and swarms of rock or pebble sized satellites forming a ring system.

Similarly, the Solar System has its share of smaller bodies. The four small inner planets are surrounded by the *Asteroid Belt,* a swarm of minor planets and rocky bodies. The four large outer planets are surrounded by the *Kuiper Belt,* a similar swarm of icy minor planets (Pluto being the largest) and the more distant *Oort Cloud.* The latter is believed to be the source of comets that fall in long orbits toward the Sun. So far out do they go, that their orbital periods are measured in tens of thousands of years.

Our Sun, complete with its retinue of planets and minor bodies, is also on the move. Just as we orbit the Sun, so *the Sun orbits the Galaxy.* We have already described something of its motion about the centre of the Galaxy (in

Chapter 1). In place of the mass of the Sun being concentrated in the centre of the Solar System, the stars in the Galaxy concentrate their mass toward its centre. The stars and other constituents of the Galaxy orbit that centre. Stars in the extended disk of the Galaxy – like our Sun – follow around in more or less circular orbits.

The stars move around the Galaxy in the same fashion as the planets move around the Sun – those further out take longer, those closer in take less time. The scale is very different, since the Galaxy is almost 100 million times bigger than the Solar System. The radius of the Sun's orbit around the Galaxy is nearly two billion times larger than the radius of the Earth's orbit about the Sun. And so the time to orbit once is larger still – about 200 million years for the Sun to go once around the Galaxy.

You may not have felt the Earth whisking along at 30 kilometres per second around the Sun. You are even less likely to feel the whole Solar System moving – not at 30, but at 230 kilometres per second, as it moves around the Galaxy! The scale removes it from human experience. The Solar System was different – we counted our orbits around the Sun with birthday candles. The year that it takes the Earth to go around the Sun is a convenient measure of human time. But *200 million years* is beyond practicality, and practically beyond imagination.

Like an army on the move, with everyone having to march together, our neighbouring stars share our motion around the Galaxy. The chief reason why those stars, when seen in the night sky, hold fixed patterns, is that we are almost all moving along together. But, like a somewhat ragged army, each soldier is not moving in perfect unison with his fellows. Over and above the general motion, our Sun has a small individual motion. But it is hardly noticeable – 20 kilometres per second in the direction of the constellation of Cygnus. If you did watch, very very patiently, then you might observe that the stars in the night sky were ever so slowly changing their relative positions. If you looked toward Cygnus, you would see them opening up their patterns as our Sun streams in that direction – the same effect that you commonly see as a screensaver on a computer. But nowhere as fast. You might have to watch patiently for perhaps a hundred thousand years!

Sadly, our time scale is all too human. In the space of a lifetime, you cannot see the Galaxy turn, any more than you could witness the growth of a tree during a five-minute coffee break. The Galaxy appears to us frozen in motion. But the *speed* of rotation is measurable. The same Doppler shifting technique used to find the speeds of recession of galaxies (Chapter 3) enables us to measure the speeds of the Galaxy's constituents. In this way, we have been able to establish the manner in which the Galaxy rotates (see Figure 6.3).

The same is true of other galaxies. Their frequent spiral shapes suggest that they too spin. Again, the rotational speeds can be measured. And even if we

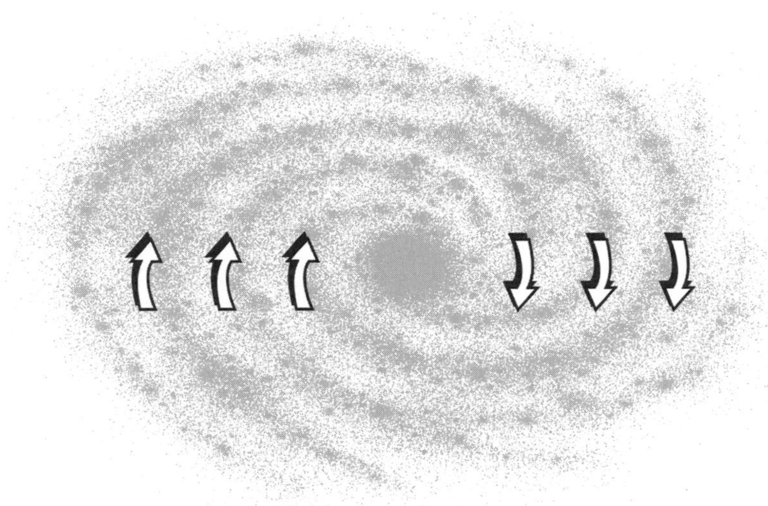

6.3 – The stars, in the disk of our Galaxy, move at more or less the same speed in their orbits about the centre. Those closer in, having a shorter distance to travel, complete an orbit in a shorter time than those further out.

cannot see the real thing, nowadays we can witness the behavior of turning galaxies by means of computer simulations. As computers have become more powerful, we have made great progress in understanding how galaxies rotate – and it is far from simple.

The chief problem has been trying to understand spiral structure. Given that the inner part of a galaxy has to go around faster than the outer part, spiral arms ought to get wound tighter and tighter. But they do not! We now realise that the stars and gas clouds in the disks of spiral galaxies go around at one speed, whereas the spiral pattern goes around at a *different speed*. Stars and gas clouds do not stay fixed in the same spiral arm, but rather pass from one arm to another. The spiral pattern is a *great wave pattern*, that builds up in the Galaxy, and swirls and agitates the gas in the Galaxy – much like the waves on water that disturb floating seagulls.

Motions within motions... within motions. Not only are the planets in motion around our Sun, our Sun and other stars in motion around the Galaxy, *but galaxies themselves are in motion*. We have already dealt with their overall *cosmological* motion, due to the expansion of the Universe (in Chapter 3). Remember the cosmological fruit cake. We also noted then that the expansion of the Universe is only apparent on large scales – even beyond groups or clusters of galaxies. Galaxies do not expand, groups of galaxies do not expand – because gravity has already arrested and reversed that expansion.

Take our Galaxy and its big neighbour, the Great Galaxy in Andromeda.

Years back, the two galaxies were driven apart by the expansion of the Universe. But the mutual gravitational pull, between the two galaxies, has since been putting on the brakes. So much so, that the Great Galaxy in Andromeda is no longer moving away from us. Quite the opposite. It's coming toward us! It comes a light year closer every four thousand years. As the distance between the two big galaxies decreases, gravity will get stronger, and the two galaxies will accelerate toward each other. It sounds like bad news, but it is going to be several billion years before the galaxies get very close, and probably each has enough sideways motion that they will not collide but instead swing around each other.

Such motion must be typical of galaxies in pairings, or groups, such as ours. But the motion of galaxies *in clusters* must be even more vigourous. Within a cluster of galaxies, the mutual mass of the membership must cause galaxies to fall into the centre of the cluster and, collisions not withstanding, out again. Galaxies must pass in and out through the centre on orbits.

Like the turning of galaxies, we see the galaxies in clusters frozen in motion. We can measure their motions in a line of sight. Some are indeed falling in, others are climbing out from the centre. Since the configurations of galaxies – that provide the mutual mass – are constantly changing, so are their orbits. Individual galaxies are likely to be following slightly different paths each time, but on average, it is a shuttling to and fro through the centre of the cluster.

You might expect this to be the end of the story. After all, any motions on a larger scale, if they exist, ought to be far milder still. Indeed, that was the thinking of the world of science, up to the 1970s. But all that changed – thanks to the Cosmic Egg. It was the Cosmic Egg that revealed even greater motion in the Universe, and *the greatest motion in which we participate.*

Even Einstein would have been amazed. In his quest to understand the way that space and time functioned (about which much more will be said in Chapter 8), he could never pinpoint *absolute rest.* Earlier hopes for the existence of a universal "ether" – which lay at rest, and through which everything moved – had been dashed by a classic experiment, in the mid nineteenth century. Every motion was *relative* to something or other, and so Einstein hit on the name "*Relativity*" to describe his work. To him, there seemed to be no absolute frame of reference at rest.

And there isn't. Because the Universe, like the giant cosmological fruitcake that it is, is expanding. Earlier we touched on the problem in pinpointing "at rest" (in Chapter 3); every galaxy in the Universe – every currant in the cake – thinks of itself at rest, and sees everything else moving. Consequently, we used to think we lived in Einstein's "everything-is-relative" world.

But the Cosmic Egg changed all that. It doesn't pinpoint absolute rest – because in an expanding universe, there isn't such a thing. But what it does allow us to find is a *local standard of rest.* This is the dough of the cosmological cake. If our Galaxy really resembled a well-behaved currant in the cake, then obviously it should not itself move, but ought to stay embedded in the texture of the cake. Such passive behavior would then mean that our Galaxy is *stationary* according to the local standard of rest.

But it isn't. Our Galaxy is moving, and so are those around it, as though the galaxies were less like currants, but more like insects that could crawl around inside the fruit cake. The motion is revealed by the Cosmic Egg.

We have already appreciated (in Chapter 2) that the inside surface is almost perfectly uniform in appearance. Only with extremely precise measurements, in the late 1980s and early 1990s, were the faintest of fluctuations – the snapshot of the early Universe – revealed. However, well before that, in the 1970s, a different fluctuation was seen: One side of the Egg appeared some 0.1 percent brighter than the other side. Were it possible to see the radiation from the Cosmic Egg by human eye, it would still be too subtle a difference to notice. But it is comfortably measurable.

The interpretation is not that the Egg is really brighter on one side. Rather, it followed another of Einstein's predictions. If we, the observers, were not to be stationary (according to the local standard of rest) but moving, then the Egg would be a bit brighter in the direction we are moving, and it would be a bit fainter in the direction we were coming from. In short, *it's us doing the moving.*

The 0.1 percent difference, between one side and the other, translates to a speed of *600 kilometres per second.* That is faster than any of the motion we have previously described – and we cannot even feel it! In the time that you have taken to read this chapter, the Earth has moved probably a million kilometres!

It is our whole Galaxy that is in motion. And more. Since the relative motions of nearby galaxies – the Local Group and vicinity – are nowhere near 600 kilometres per second, it can only be that our surrounding galaxies share that motion. We are, all of us, on the move – an astonishing streaming motion (as suggested in Figure 6.4).

This discovery overthrew the notion that the galaxies are passive and as stationary as the currants in the cake. It showed that motions occur in the Universe on previously unimagined scales.

The direction of our motion is roughly that toward the constellation of Centaurus in the southern sky. This is merely to suggest direction – the stars that make up the constellation are, of course, within our Galaxy, completely in the foreground. It is not immediately obvious why we should be headed in that direction, or to where we are going.

In retrospect, a part of the motion might have been anticipated. So far, we have seen that all motion is caused by gravity, and for gravity to act, there has to be mass. What concentrations of mass outside our local group of galaxies

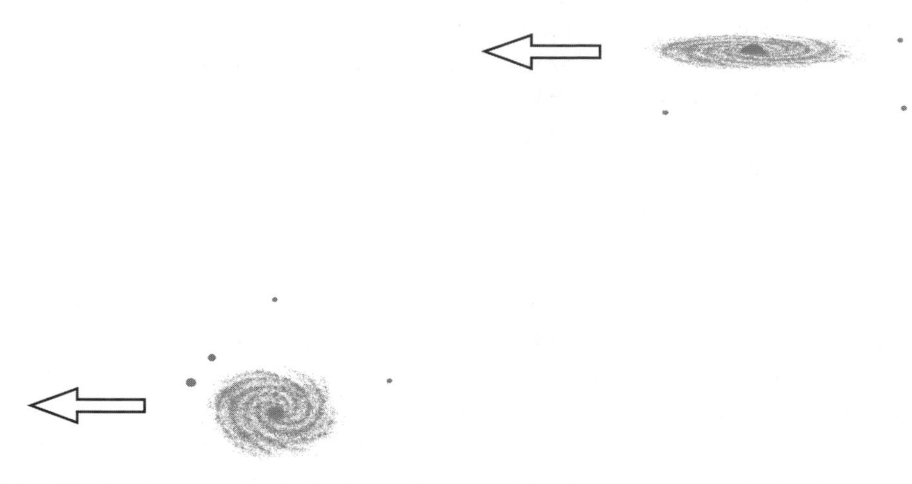

6.4 The local group of galaxies is moving at 600 km/s, compared to the local standard of rest.

could be responsible? An obvious candidate would be the nearest cluster of galaxies, the Virgo Cluster. But the Virgo Cluster lies about 44 degrees away from the direction we are headed. Something else is pulling us.

By the late 1980s, two different techniques offered insight as to where we might be headed. Both measure distances of galaxies, independently of their cosmological velocities. Both focussed on internal motion within galaxies – these indicated how much mass a Galaxy had, and therefore how much luminosity it might produce. They were complementary. One, known to researchers as the "modified Faber-Jackson relation" applied only to elliptical galaxies. The other, the "Tully-Fisher" relation works with spiral galaxies. Though the fine-tuning of these relations continued for years afterwards, they have generally been very reliable – particularly when applied to groups, rather than to individual galaxies.

If we know the distance of a galaxy, we can predict its *cosmological velocity* – due to the expansion of the universe. If we compare that to the observed velocity of the galaxy – then the difference must reflect the actual motion of the galaxy, toward or away from us, independent of the expansion of the Universe. In this way, researchers have been able to map the large-scale streaming motions of galaxies.

In 1986, a group of researchers, dubbed the "Seven Samurai", caused a sensation when they announced that their work with elliptical galaxies not only confirmed the general motion shown by the Cosmic Egg, but indicated that the flow was converging on a point some 200 million light years out in the Centaurus direction – an apparent mass concentration they christened the

Great Attractor. Whatever it was, it seemed to be pulling all the galaxies in the surrounding Universe toward it.

We may be heading for the Great Attractor, but we will never reach it. Due to the expansion of the Universe, the supposed Great Attractor region is moving away at 4500 kilometres per second. Our Galaxy is more like a small piece of paper that is sucked along by, and seems to chase, an Intercity train – but never manages to catch it up!

The discovery of a Great Attractor came at a time when large-scale structures (the topic of Chapter 4) were just beginning to be mapped. The Great Attractor region showed up as a concentration of galaxies in the direction of the constellation of Centaurus. It would appear to lie in the heart of the Centaurus Wall (see Map 5b in the colour section *The Universe around us*).

In the years that have followed, further observations have accumulated, and the role of the Great Attractor debated. The new data generally confirms the convergence on the Great Attractor region, but suggests that it is not entirely responsible for the 600 kilometres per second. Other flows have been detected, and the overall picture may be best summarised by a modified version of our earlier map of nearby large-scale structures, that now includes recent velocity measurements – see Figure 6.5.

These localised flows still leave the possibility of there being a general streaming of galaxies on an even larger scale. Notice, in the diagram, a tendency of flow toward the upper left. Currently, there is much discussion about using distant clusters of galaxies as a reference for such motions, following an early controversial result that they too might be streaming. However, the evidence is still tentative. Perhaps only time will tell – to how big a scale the motions in the Universe extend.

And so we see the contents of our Universe are constantly on the move. *Gravity* appears responsible. It is probably the only physical force that governs the Universe on its larger scales. If we are to understand the cosmos, we have to understand gravity.

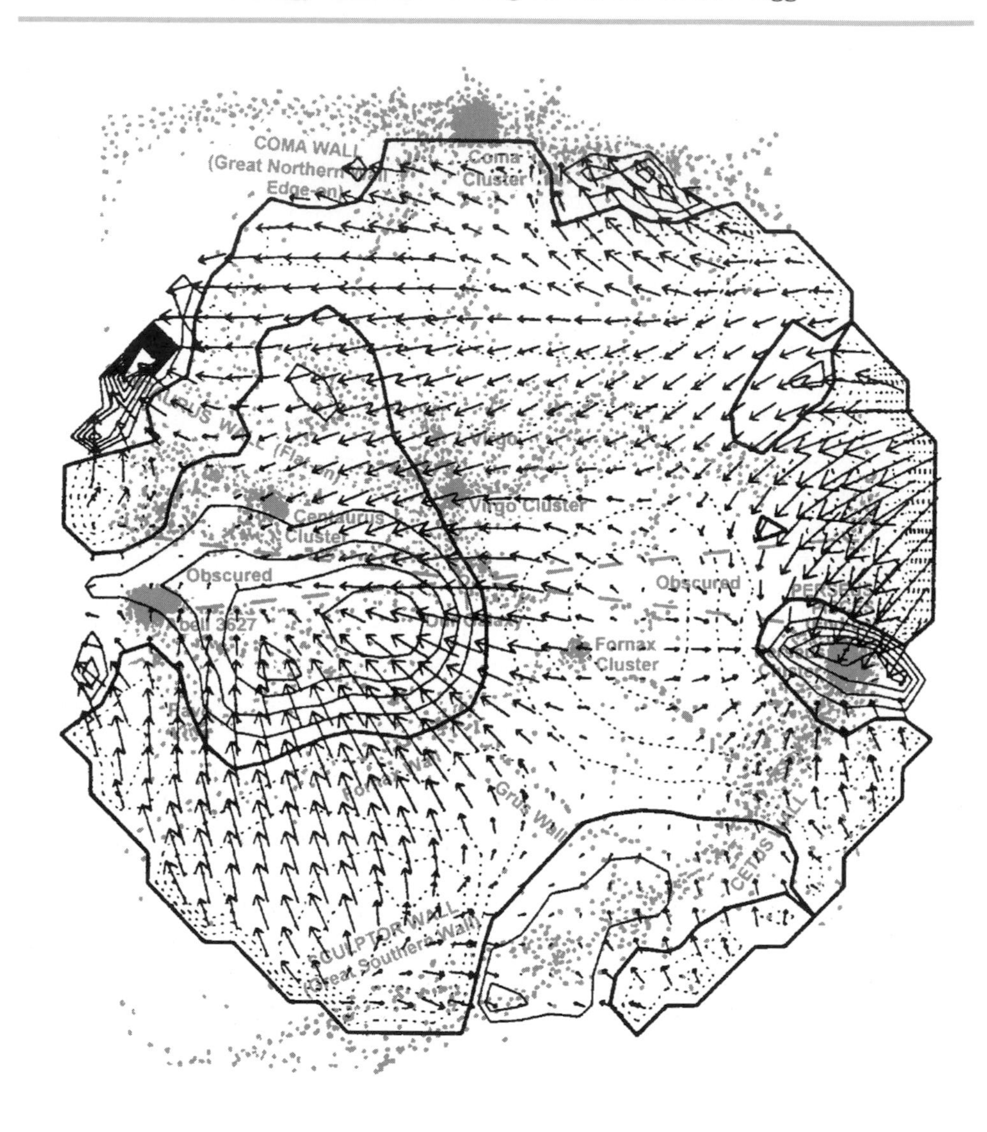

6.5 The arrows show streaming motions of galaxies in relation to our map of nearby large-scale structures. The convergence toward the Great Attractor region (left of centre) is apparent; similarly there are flows toward Perseus-Pisces (right) and the Coma Wall (top). (Adapted from a map by L.N. da Costa and collaborators.)

Chapter Seven

Gravity and Antigravity

Gravity gets you down. Sometimes it's a bugbear, but usually it's a convenience. It keeps coffee in cups, cars on roads, trains on tracks, and your feet on the ground – to mention but a few benefits. Without gravity, the spin of the Earth would, in any case, hurl us off into space.

The Universe wouldn't be the same without it. Without gravity, no stars would form or shine. Without gravity, no galaxies would assemble. Without gravity, Earth could not offer a home to humans. In that sense, we could not be here were it not for gravity.

Nevertheless, we humans can sometimes get along without it – as far as our physiology is concerned. For instance, some of the cosmonauts in the Mir space station clocked up more than a year in space – living in weightlessness. There were some minor problems. In space, the circulatory system tends to push too much blood to the head, and too little to the legs, because it is accustomed to working against gravity. In the long term, muscles diminish unless constantly exercised, since their strength is not needed when floating in space. Also bones lose calcium, because the human skeleton need not be so strong. Down on Earth, our bodies are designed to withstand gravity. We are so familiar with it on an everyday scale.

The Universe is all too familiar with it on a cosmic scale. Gravity has governed its evolution and behaviour. It has changed the way in which matter is distributed. Like raindrops that condense in a cloud, gravity has turned an almost smooth spread of gas into the various astronomical bodies -from minor planets right the way up to large-scale structures.

It also sets everything in motion (the topic of the previous chapter). But, had this book been written a few hundred years back, I might have tried spinning a very different story. Gravity then was thought to apply *only* down here on Earth. The heavens were above – because gravity didn't apply up there. Rather the celestial vault was considered to be the realm of perfection – immune to Earthy matters. Angels flew and crystalline spheres spun with uniform regularity, without being hassled by gravity.

One man changed all that. Isaac Newton. Raised in Lincolnshire, he followed an academic career at Trinity College, Cambridge. However, his studies were interrupted by the Great Plague, which closed the university and Newton, like many, sought rural refuge back on the family farm.

Legend has it that it all happened one afternoon (in 1665 or 1666) when Newton was sitting outside the farmhouse, under an apple tree. An apple fell from a tree on to Newton's head, whereupon he said:

$$F = \frac{Gm_1m_2}{r^2}$$

He probably said a few other things as well, but happily history does not record them. The formula above is Newton's famous Law of Gravity. It gives the gravitational force (F) between two masses (m_1 and m_2) separated by a distance (r), (where G is a constant). It means the apple pulls on the Earth, just as much as the Earth pulls on the apple. Not only then did the apple fall, the Earth, and Newton's head, came up to meet it! However, given their relative masses, the movement of the Earth was negligible.

Newton had the mathematical ability to apply this law to the motion of planets around the Sun, and found that it explained Kepler's Laws – a set of rules as to how the planets move around the Sun. In other words, Kepler had shown how, but Newton explained why.

The world might have been still ignorant of Newton's breakthrough, had his friend Edmund Halley – later famous for his comet – not urged and even financed Newton to publish his results. Newton's *Principia* is the basis for the understanding as to how things move in the universe, from pendulum clocks to roller coaster rides, to planets and beyond. It is the foundation of all technology that involves force and movement.

Halley was able to use Newton's Law to examine the way that comets moved in the Solar System. Comets have long been seen as ghostly apparitions that appeared on occasions in the night sky. Halley removed much of their mystical character by showing that they orbited around the Sun like planets, but their orbits were much more elliptical. He found that three of the comets, seen back in history, all had the same orbit, and the same interval between passing around the Sun. So he concluded they were all the same comet, and predicted when it would next return. Sadly, he did not live long enough to see his prediction come true, but that famous comet bears his name. In truth, a comet

The Universe around us, in three dimensions

The scenes in this section may be viewed with the ChromoDepth(tm) spectacles found at the back of this book.

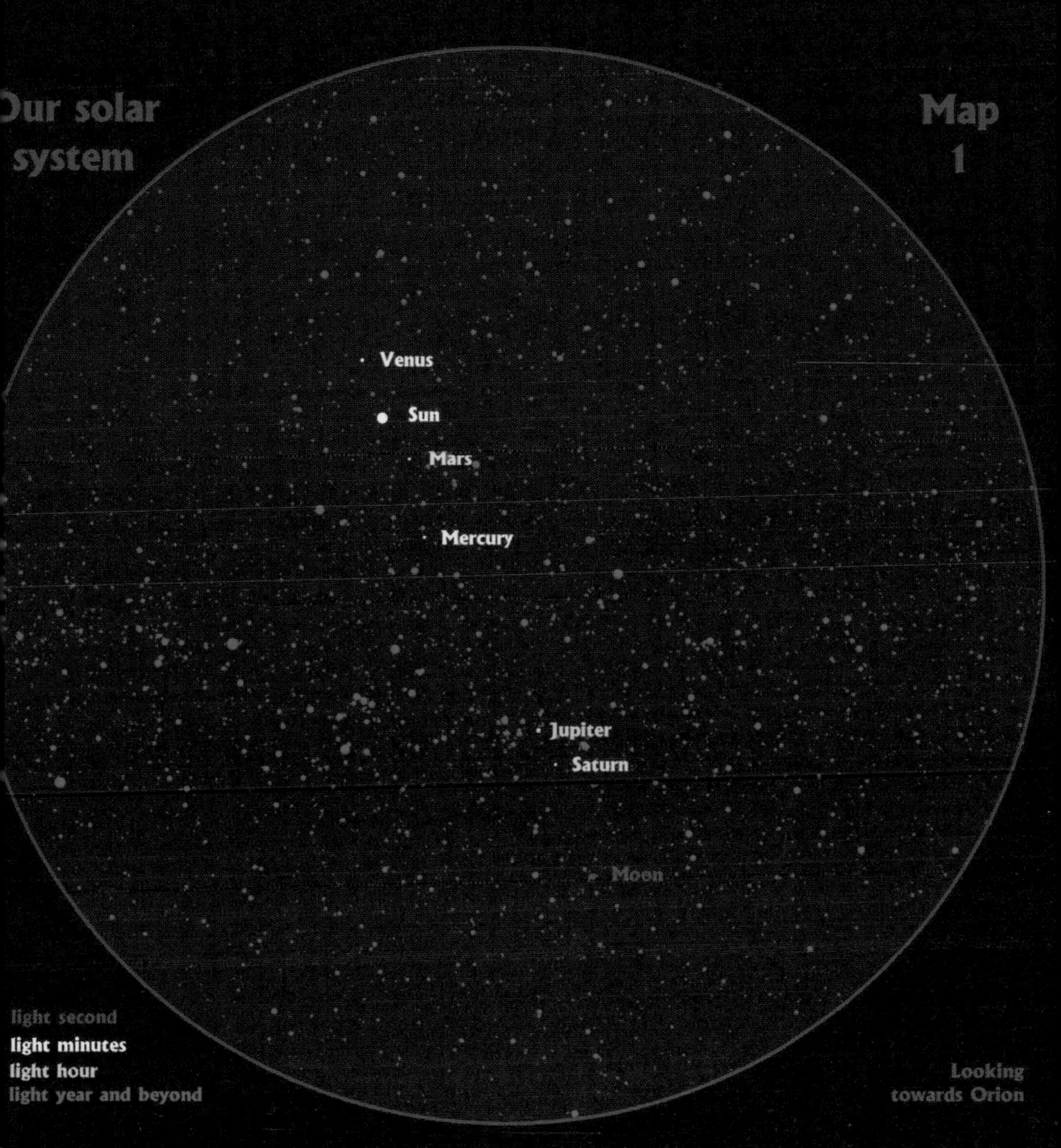

The colour diagrams in this section consist of a sequence of diagrams portraying the universe we see around us to ever greater depth. This first diagram shows the nearby objects of the Solar System, separated from the much more distant starry background. While the Sun may not normally be seen against the other stars, the nearby planets are often confused with bright stars. The scene is set at a time when the Sun, Moon and five naked-eye planets all lay in the same part of the sky (late July 2000). (See Acknowledgements for detailed credits and databases used here.)

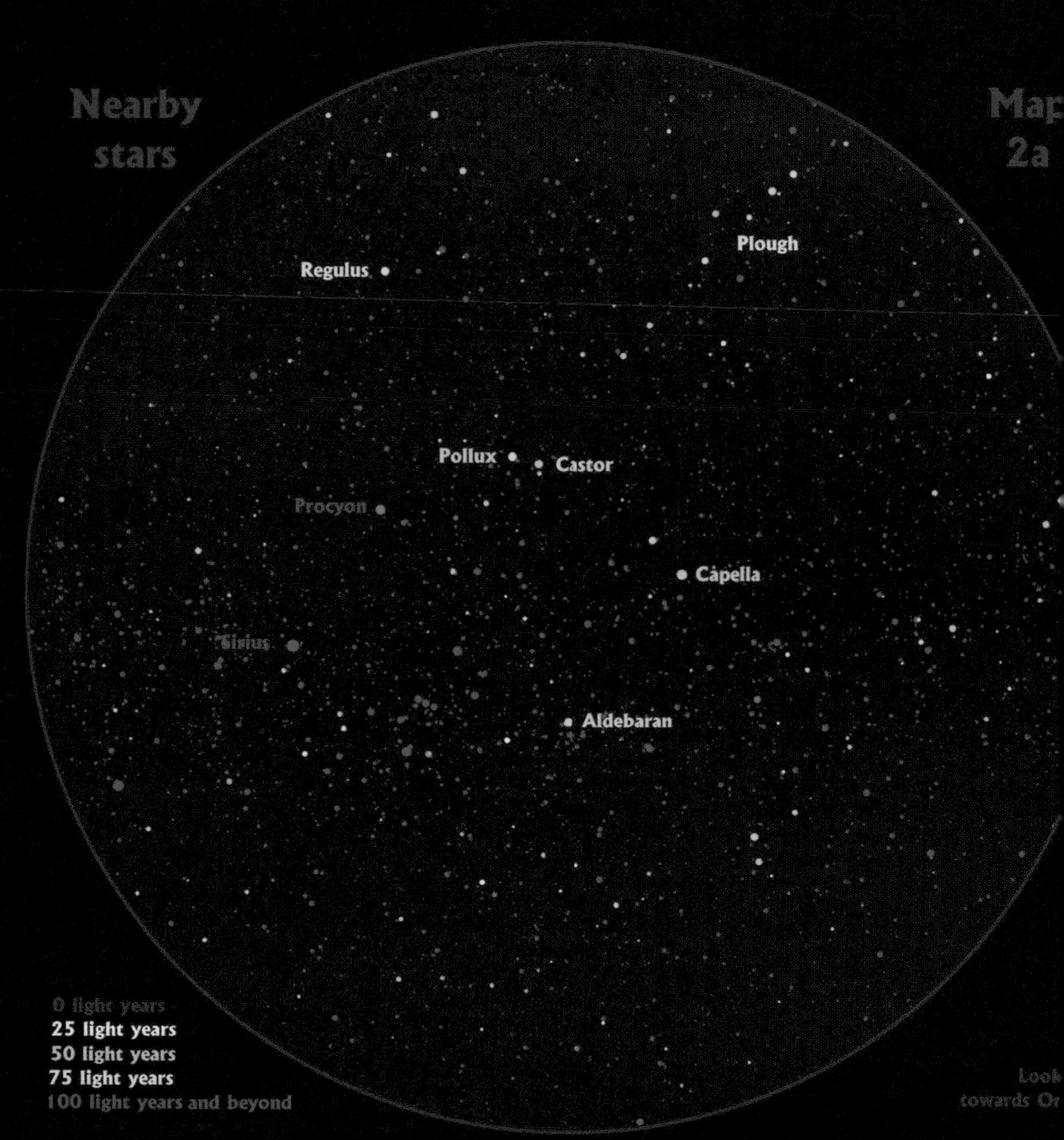

In these diagrams, *distance* is conveyed by *colour*. The closest objects are shown in red, the most distant i blue. The colours of the spectrum between red and blue indicate intermediate distances. For persons wit normal colour vision, the ChromoDepth(tm) spectacles will transform colour into stereoscopic depth. I the pair of maps here, the colour range is spread from zero to a hundred light years to make the neares stars obvious. The two maps cover the entire sky; for clarity, the objects in our Solar System (in Map 1 have been omitted.

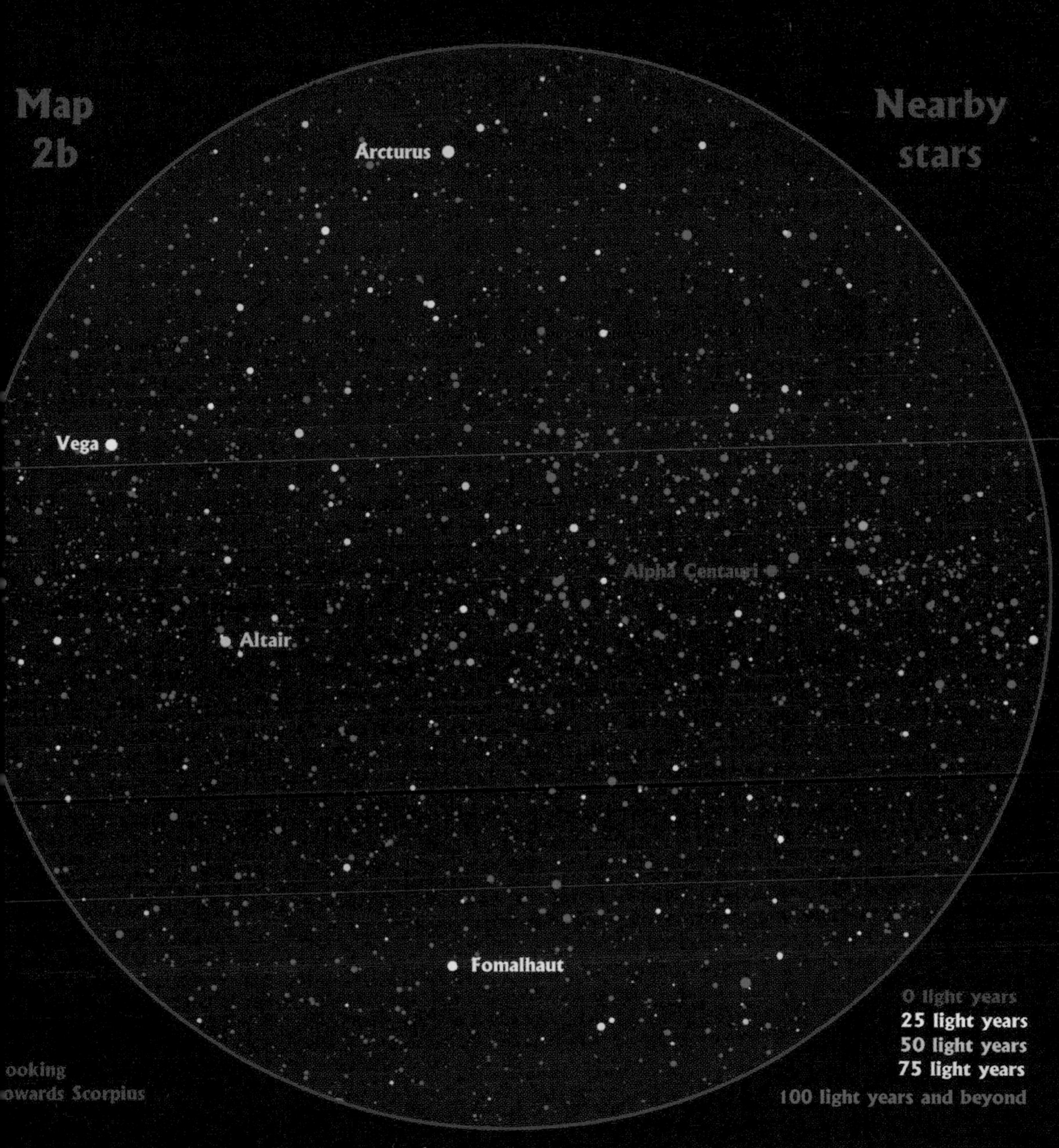

'he stars labelled in these maps are amongst the brightest stars visible in the night sky. The closest is lpha Centauri (right of centre in the map above). Other very nearby stars are Sirius (brightest in the night ky) and Procyon (both on the opposite page). Stars, however, show a great variation in their actual ıminosity; one can see many faint nearby stars, while other stars that appear bright in the night sky are till more distant that those highlighted here (all stars more distant than 100 light years are shown as deep lue).

The colour coding now spreads from nought to a thousand light years, to accommodate almost all the stars visible to the naked eye. As before, this pair of maps covers the entire sky and foreground objects in our solar system have been omitted. Canopus (the second brightest star in the night sky), many of the stars of Orion, Spica and Archernar (both opposite) are examples of high-luminosity stars. Two of the nearest star clusters are apparent – the Hyades (behind the foreground star Aldebaran in Map 2a) and the compact Pleiades.

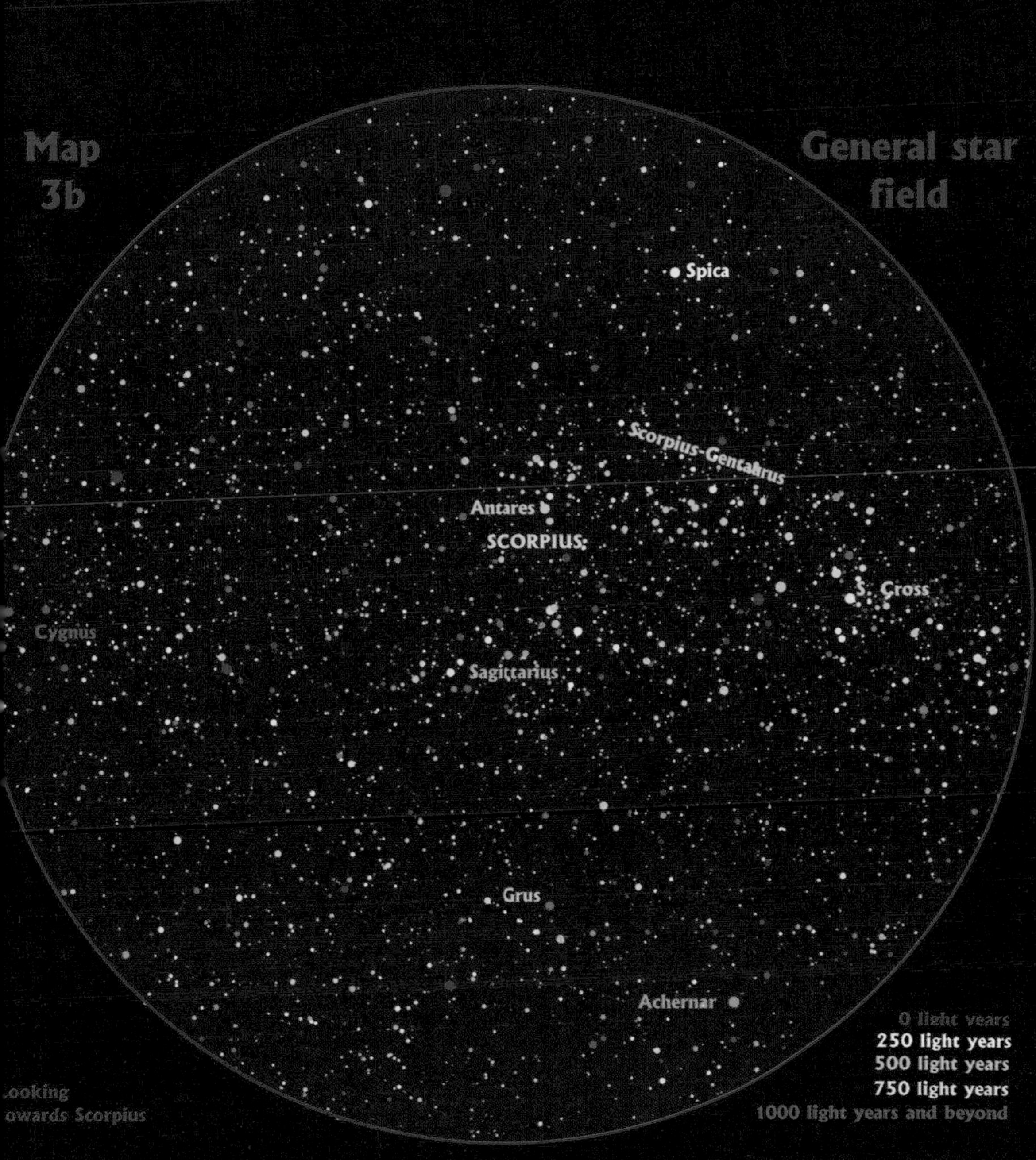

Many constellations – apparent groupings of bright stars – are disrupted by the separation in distance, and are much more difficult to recognise when seen in three dimensions than in two. True groupings of stars can be seen far easier in three dimensions, especially the Scorpius-Centaurus Association, which includes Antares (the brightest star in the constellation of Scorpius) and stretches over a wide angle in the night sky. This association represents relatively young stars formed on the inner edge of our local spiral arm in the Galaxy.

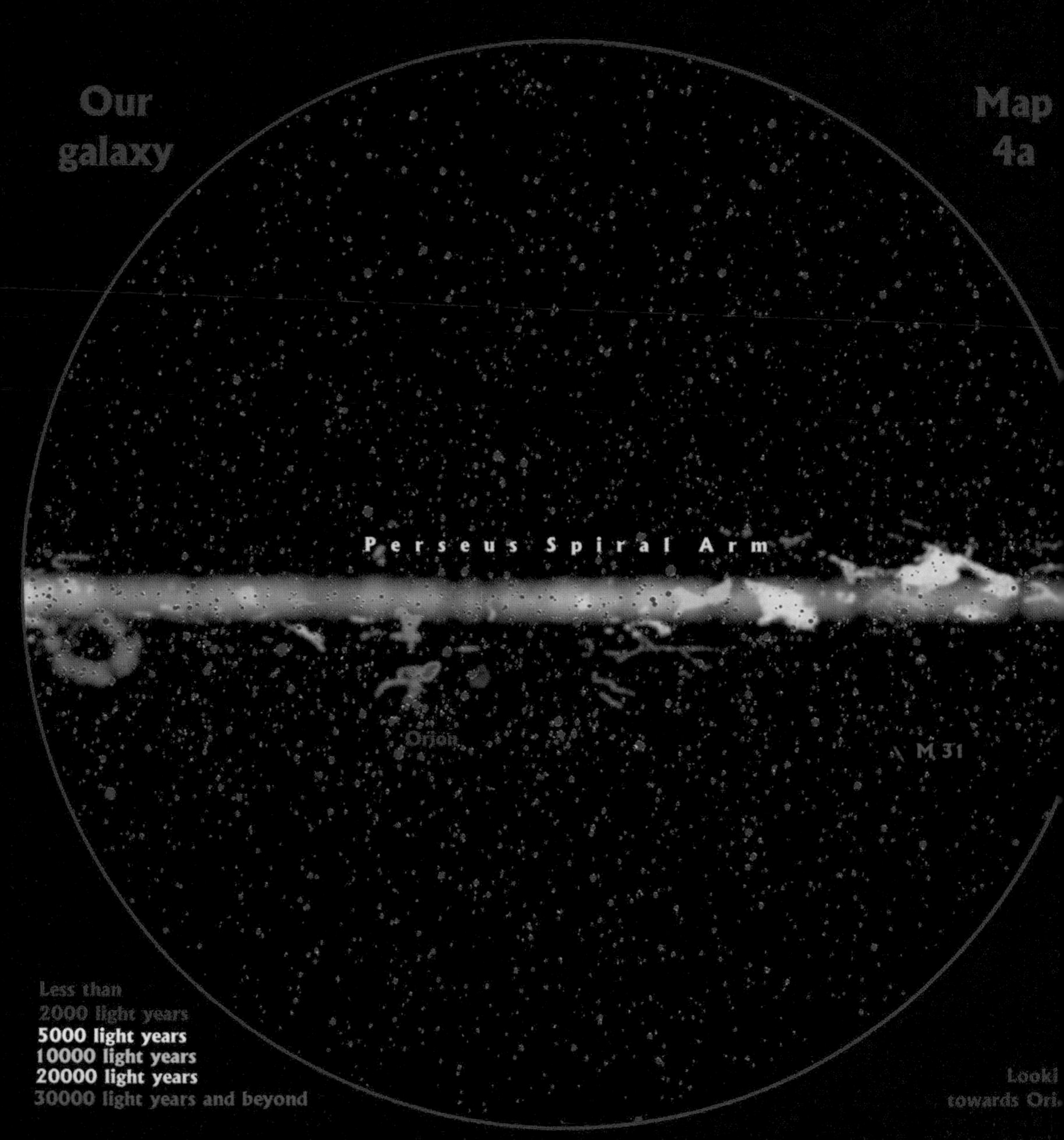

The galaxy is a great "city" of stars and interstellar clouds (Figure 1.3 on page 7 shows its probable external appearance). Our Solar system is situated far from its centre. Lying within its flattened disk, we see the Galaxy encircling us as a band, the Milky Way. The map above, shows the view looking away from the centre of the Galaxy, the map on the opposite page is the view looking towards the centre. All the stars visible individually to the naked eye now form the foreground. The colour scale now stretches to 30,000 light years, our distance from the Galaxy's centre.

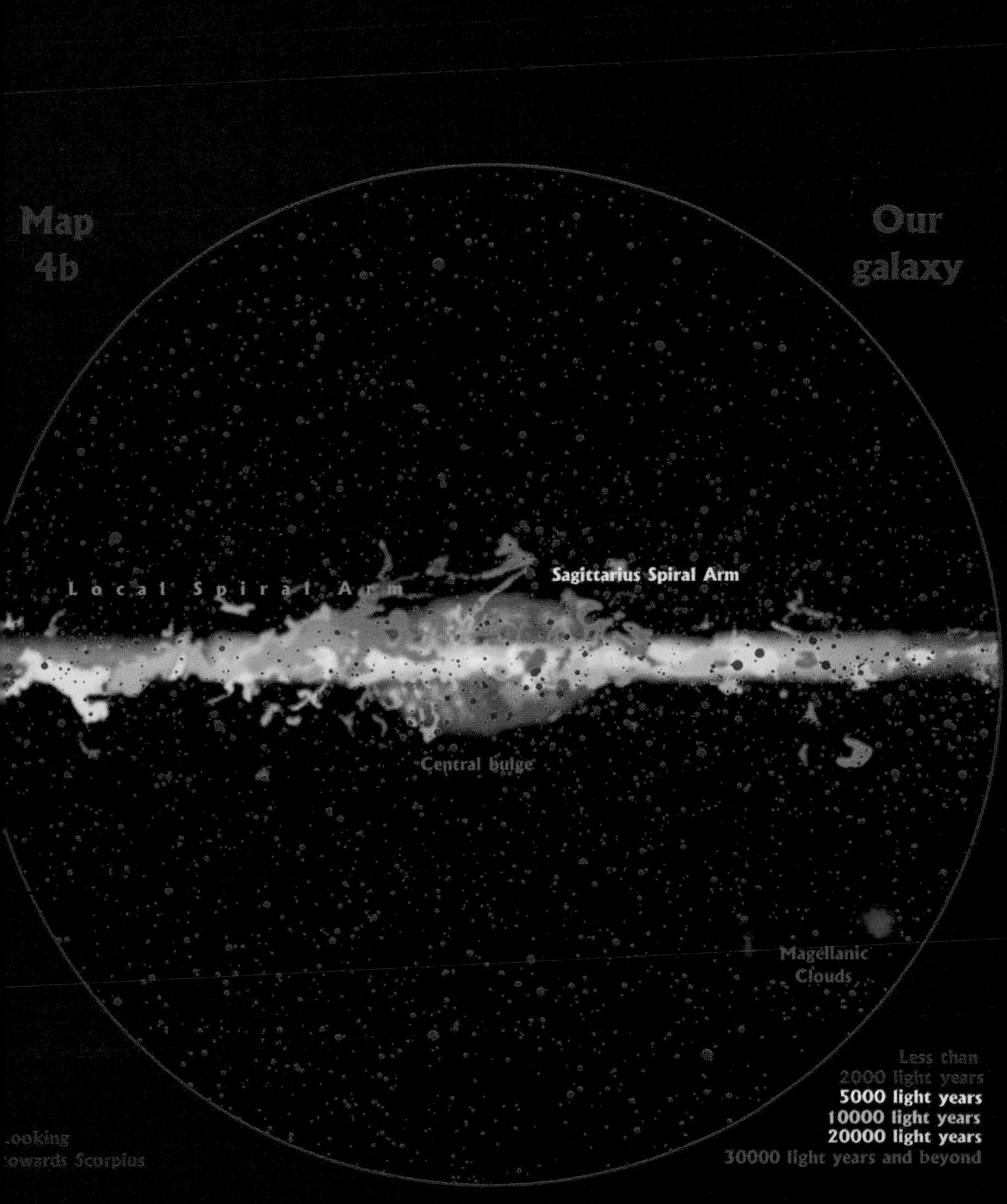

The central bulge of the Galaxy is apparent when we look in the direction of the constellations of Scorpius and Sagittarius. However our view is heavily obscured by the many dust clouds in the intervening spiral structure, especially those in the Sagittarius spiral arm, next in from our local arm. Normally such clouds would be dark and opaque (as seen in the previous colour section). Here they are depicted as "luminous", so as to be able to colour code them according to distance, The brightest neighbouring galaxies, to our own, are also shown.

Other galaxies

Map 5a

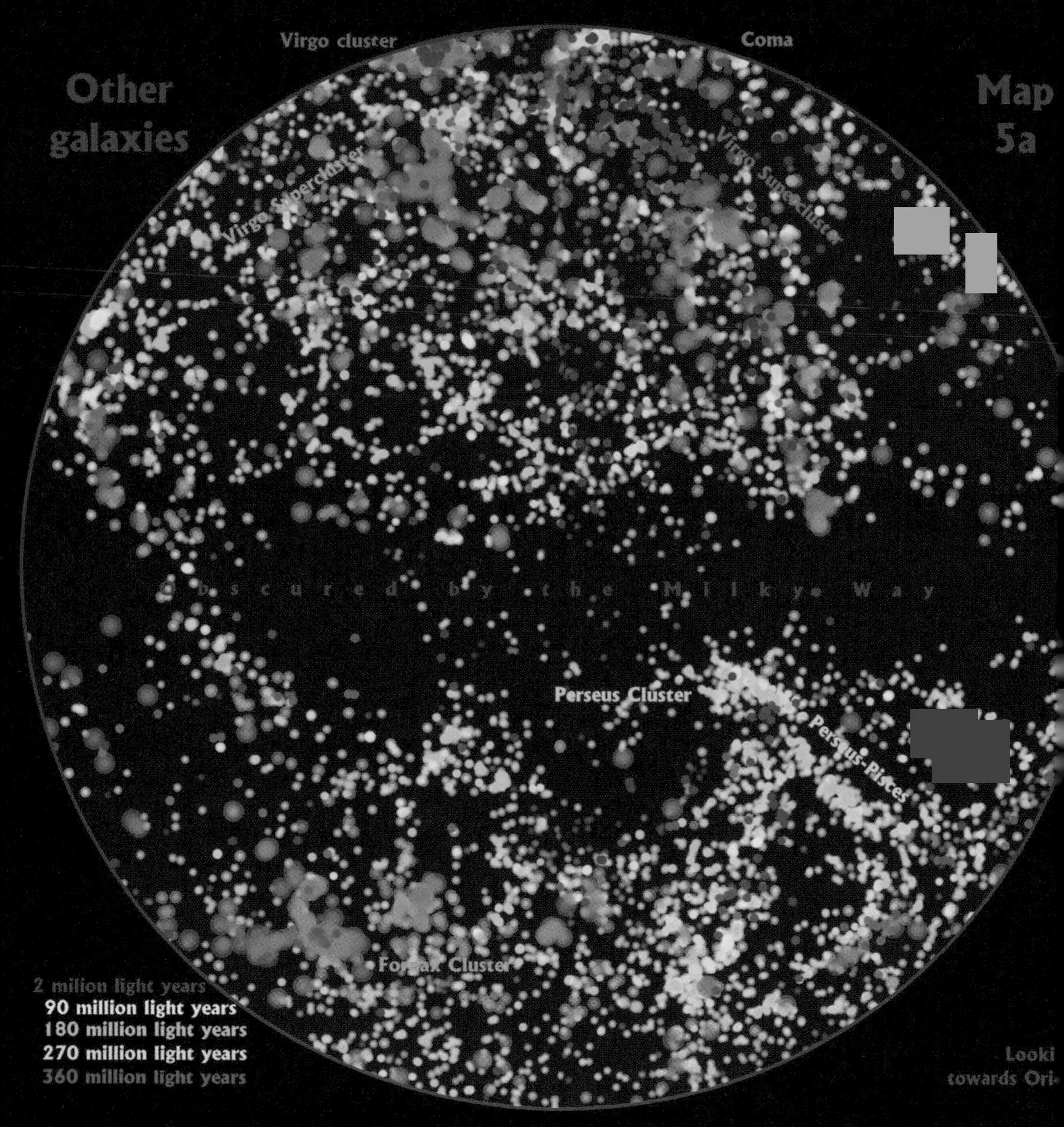

The universe is populated by many galaxies similar to our own, as our view moves beyond naked eye visibility, with our Galaxy hereafter omitted for clarity. We look to a depth 10,000 times greater than the previous view. The maps include known galaxies out to a distance of 360 million light years. The galaxies congregate into large-scale structures – a labyrinth of interconnected wall-like and ribbon-like superclusters, that exhibits a frothy texture. Numerous voids permeate the cosmos. This pair of maps shows the nearest large-scale structures.

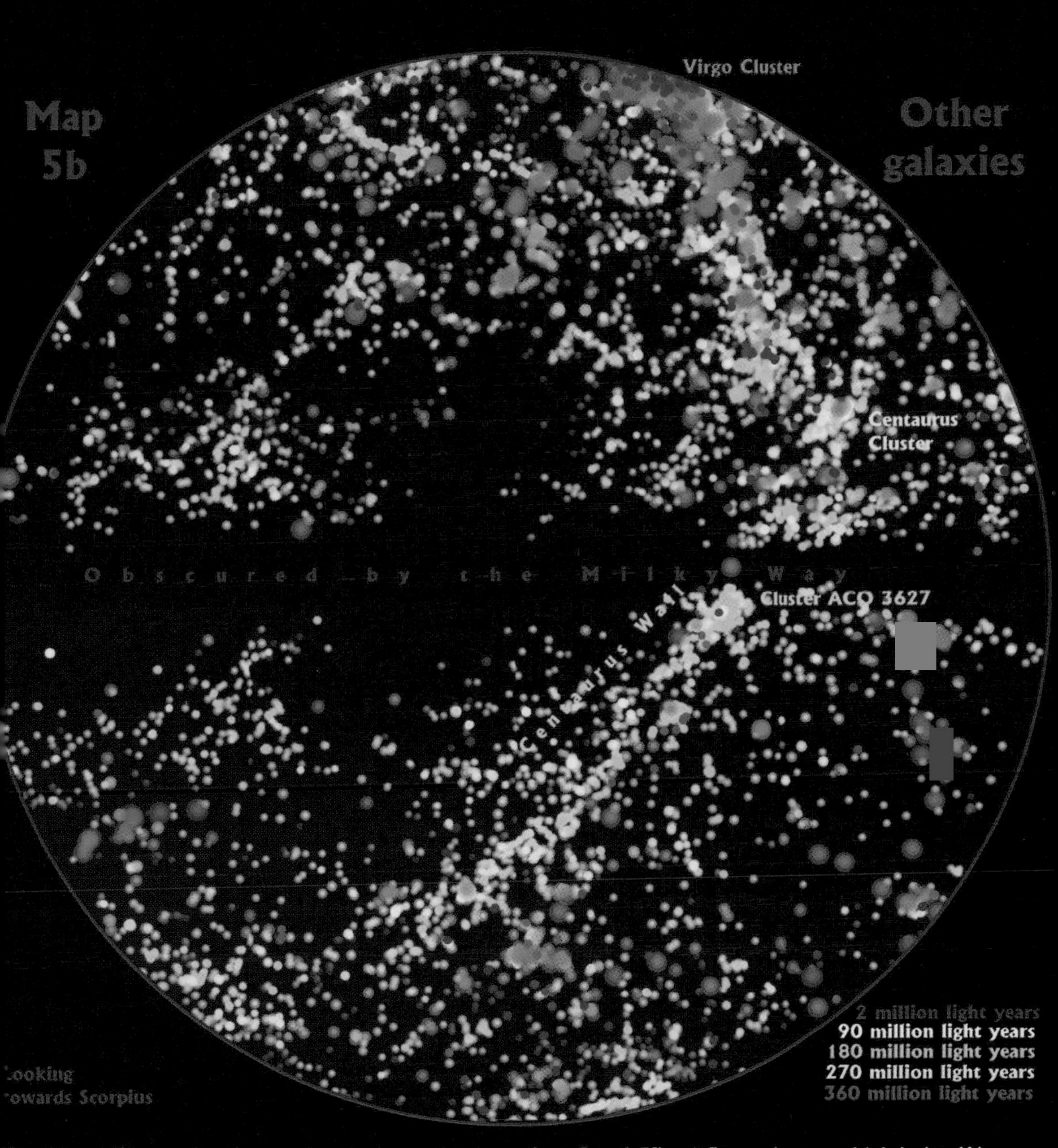

The Virgo Cluster, the nearest cluster, forms the core of our Local (Virgo) Supercluster, which in itself is a part of the more extensive Centaurus Wall. That structure appears as the most dominant feature in these maps; it includes the Centaurus Cluster and the rich cluster ACO 3627. We lie towards the edge of this massive feature. In the opposite direction is equally massive but more distant Perseus-Pisces supercluster. Still more distant is the Coma Cluster. (Figure 4.5 in the text shows a "plan" view of these structures.)

Clusters of galaxies

Map 6a

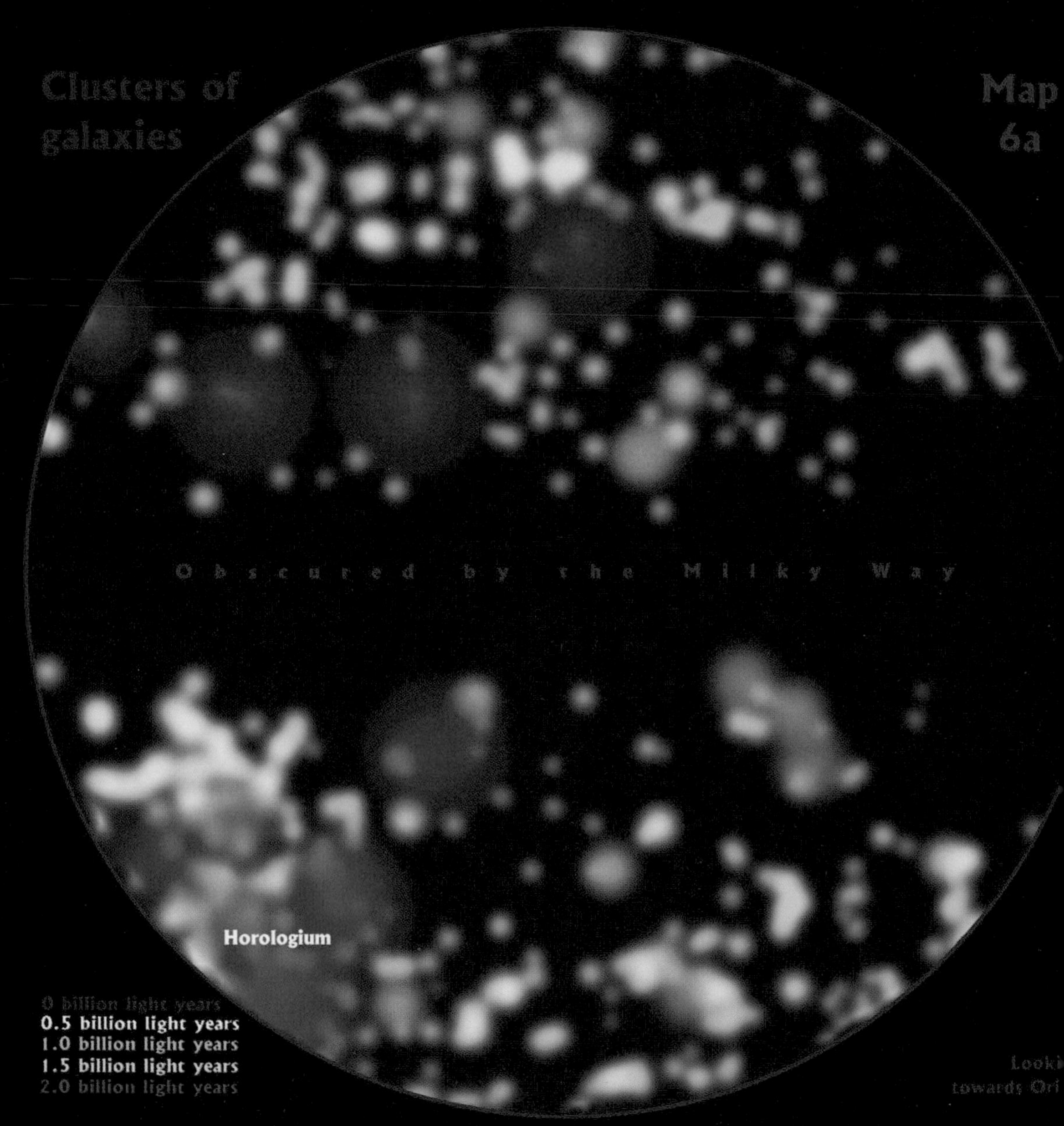

We look still deeper into the cosmos as our colour scale stretches to two billion light years. On this scale, we have yet to map galaxies all over the sky – we have so far only probed in limited directions. We no longer show individual galaxies in this view. However, we have plotted the distribution of rich clusters of galaxies (those originally identified by George Abell and collaborators). Such clusters are believed to be beacons of large-scale structures – they show us how the galaxies concentrate in the universe.

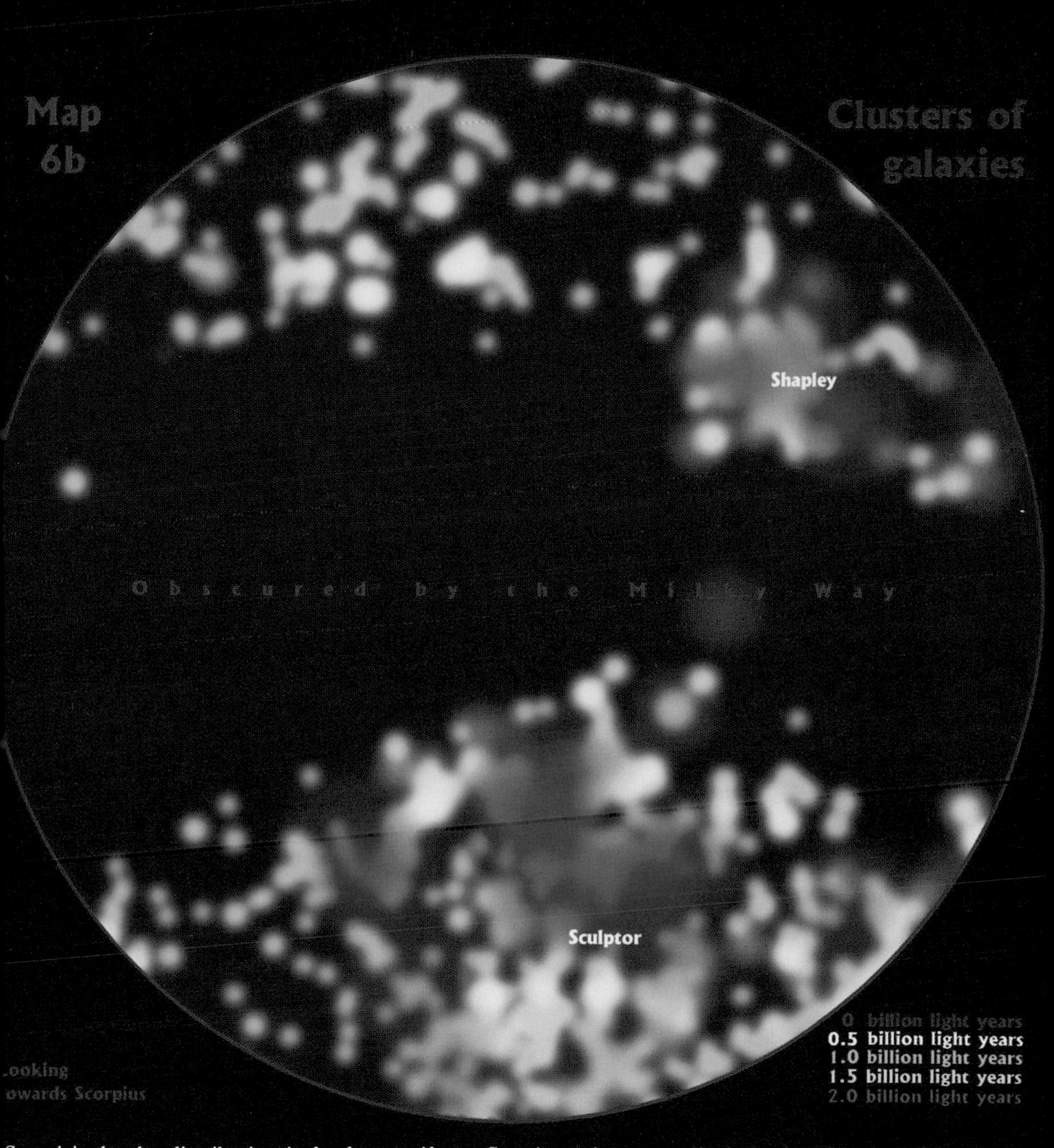

Surprisingly, the distribution is far from uniform, Certain regions – particularly the "Shapley" (after its discoverer), Horologium and Sculptor regions seem to show great concentrations, a possible but tentative indication that large-scale inhomogeneities exist on this scale in the universe. Such overdense regions disturb our local neighbourhood as their gravitational influence sets galaxies streaming in their direction. Our own Galaxy moves at some 600 km/s in the general direction towards the right hand side of the map above.

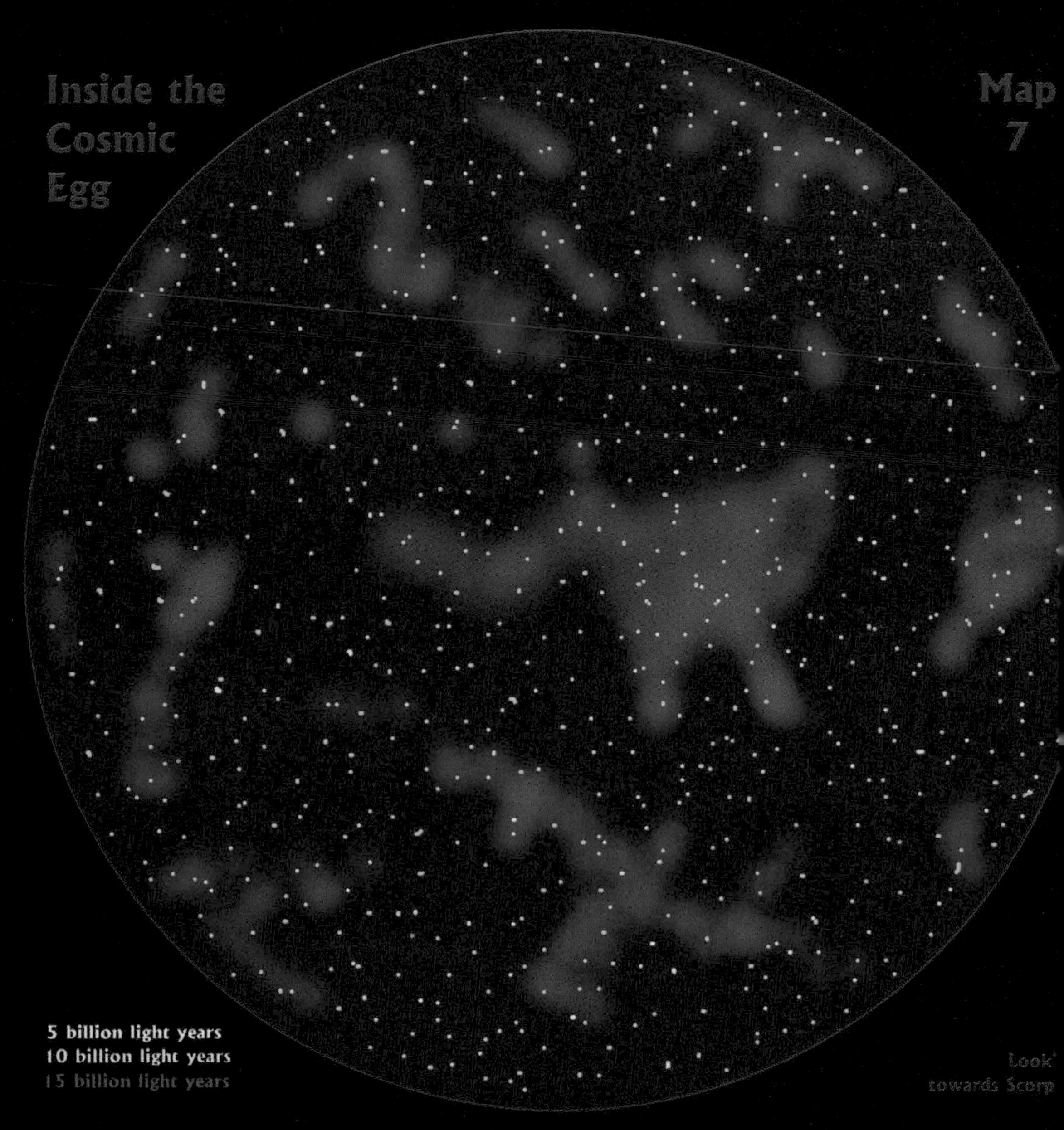

Our final map stretches to the boundary of the visible universe; the image of the early universe, captured on the inside shell of the Cosmic Egg (explained in Chapter 2) provides the background (blue). In front of it, the pointlike images represent an era of quasars. Unlike the previous maps, the representation of the quasars here is schematic, since we have no data set uniform in coverage over the sky to this depth. We have however probed in limited directions (as reported in Chapter 5). What lies beyond the limit of this map we do not know, and shall never discover.

is nothing more than a large cloud of water vapour, seeded by an icy minor planet that ventures too close to the Sun.

So comets were found to move in and out through the Solar System, literally shattering the older concept of there being a set of crystalline spheres. It was an enormous enlightenment. Moreover, the Galilean Moons of Jupiter – and other moons – all obeyed Newton's Law. Instead of there being a separate heavenly realm above, what happened in the heavens above was governed by the same everyday law that operated on Earth below.

Much later, in the late 18th and early 19th centuries, William Herschel was to discover large numbers of double stars, and his son, John, was able to show that they too orbited around one another according to Newton's Law. Suddenly understanding of how the Universe worked lay within our grasp. It was a great intellectual achievement for the human race.

Newton had never been able to know the value of G – the constant he had used in his formula. It was first successfully measured in a laboratory by Henry Cavendish in 1798. Once it was known, the mass of the Earth, the mass of the Sun, the mass of planets, and the combined masses of double stars, could be determined by Newton's Law.

Today we use Newton's Law, with extreme precision, to direct spacecraft to all sorts of places in the Solar System. Even the slight deflection in the path of a spacecraft as it passes an asteroid is enough to determine the mass of that minor body.

We may not yet be able to send spacecraft out to other stars; we may barely perceive the movement of the other stars. Nevertheless, we can extend the use of Newton's Law way beyond the confines of our Solar System, and even other solar systems, to the motion of celestial bodies in our Galaxy.

We mentioned, in previous chapters, something of the complexities with which our Galaxy rotates, as to how our Sun and its family of planets orbit around the Galaxy. There is, however, a mystery as to how spiral galaxies rotate. If most of the mass of a galaxy is concentrated at its centre, then the stars in the disk ought to go around in the same fashion as the planets go around our Sun. The further a planet is from our Sun, the slower its orbital velocity. Surprisingly, the orbital velocities of the stars going around galaxies do not diminish – but seem to remain more or less the same as far out as we can measure.

Are we seeing Newton's Law of Gravity break down? Are we trying to extend it, unreasonably, to a scale millions of times larger than the scale of the Solar System? Such is the confidence that the scientific community has gained in the use of Newton's Law, that this is considered unlikely. Moreover, Newton's Law has been applied on an even larger scale – to simulate near collisions between galaxies – and was found to give accurate results.

Rather, the general consensus is that much of the Galaxy's mass is not concentrated in the centre. Yet, the *light* from the Galaxy shows a central condensation. The uncomfortable conclusion is therefore that there has to be additional mass in the Galaxy, away from its centre, that emits negligible light. But what is it? We detect it gravitationally, but we cannot see it. Somewhat to our embarrassment, science does not as yet know what it could be.

Up until now, we have been working on the blithe assumption that we could see everything in the universe. But are we quite wrong? After all, everyday life proves otherwise. When flying at night in an aircraft, you might look down and see cities illuminated like sparkling jewels, but the forests and mountains remain in the dark. On the same basis, there might be a lot of things in the universe that are dark and which we therefore cannot see. Is everything compelled to give off light? The indication, from the way a spiral galaxy rotates, is that there is substantial additional mass – not normal stars, and without normal luminosity. *Dark matter!*

With this in mind, let's look at the *collective* behaviour of galaxies. The focus of the discussion is how gravity acts on very large scales. It is crucial to our understanding of the Universe, and has been a key issue in cosmology for the past quarter century. Newton's Law suggests that the larger the scale, the milder the gravitational force. We might therefore expect the effect of gravity to dwindle toward insignificance – *unless* ever greater amounts of dark matter can be called into play.

To gain further perspective, consider the situation as science used to know it. Back in the 1960s, the Universe seemed a simpler place. Stars, conveniently luminous, were thought to account for most of the mass in the Universe. They assembled into galaxies, up to a million million at a time, and the galaxies were believed to rotate exactly as Newton's Law acting between the stars predicted. Galaxies occasionally formed clusters, within which they moved to and fro. And that was it. Clusters of galaxies were believed to be the largest entities in the Universe, the largest things that could be formed by gravity. To form anything larger was impossible, for gravity would require a time scale tens or hundreds of times longer. In the legacy of Edwin Hubble, galaxies outside clusters were scattered as randomly as were the stars in the night sky.

Yet that picture was wrong. As we recounted earlier (Chapter 4), galaxies outside clusters are not randomly scattered, but assembled into large-scale structures. The old ideas as to how gravity works – or what it works on – needed revision.

In fact the problem was known since the 1930s, but those who encountered the irascible character and dogmatic views of Fritz Zwicky preferred to discount his research. (The author, who worked as a student with Zwicky, has a very different perspective on this great scientist). Zwicky was an astronomer of

Swiss nationality who nevertheless spent most of his time, and pursued his career, in California (at the, then, newly founded Jet Propulsion Laboratory and the California Institute of Technology). He was constantly at odds with his colleagues, but nevertheless was reluctantly given observing time on the largest telescopes – the *100-inch* and later the *200-inch* reflectors.

Zwicky had examined the way the galaxies in the Coma Cluster were moving, and had found that they were shifting around much faster than expected, which implied that much more mass must be present. At the time Zwicky suggested that some form of dark matter must be present. Trying to understand this "missing mass" was, for years, a bothersome little problem. There were some obvious possible explanations: galaxies could be more extended, or matter could be present between galaxies.

Zwicky died in 1974, too early to see what he had started. Within the following decade, the rotations of galaxies had been explored – which showed that they must have much more dark matter within them than scientists thought. But above all, large-scale structures were recognised, previously thought to be too large for gravity to construct.

If large-scale structures developed from the incredibly small fluctuations imprinted on the inside shell of the Cosmic Egg – that early snapshot of the universe – then some mechanism brought about a complete rearrangement of the distribution of matter in the Universe. Matter had to be shifted out of the voids, and clumped into the irregular shapes of the large-scale structures. It was large-scale movement with substantial large-scale motions. It had to be the action of gravity.

Gravity cannot be turned off. Large-scale structures are still growing. Large-scale movements must still be evident. Indeed they are (as described in the previous chapter). Our Galaxy and surroundings are falling at that incredible speed of 600 kilometres per second. The *Great Attractor* seems to be largely, or at least partially, responsible. If the attraction is gravity, then the Great Attractor must be a great concentration of mass – a density of matter far far above the cosmic norm.

Can we see it? Only partially. Figure 7.1 shows the distribution of galaxies over half the sky, centred on the direction in which we are moving. A general concentration of galaxies in the *Great Attractor* region, is apparent. Yet it does not add up to anywhere near as many galaxies as one would need to get the required mass of the *Great Attractor*. Furthermore, had we shown the opposite half of the sky, a similar concentration would have shown up in the direction of *Perseus-Pisces.*

One solution would be to involve additional mass – dark matter. If the *Great Attractor* region is mainly responsible for our motion in its general direction, then most of its mass is unseen, or Newton is wrong. Once again, we must

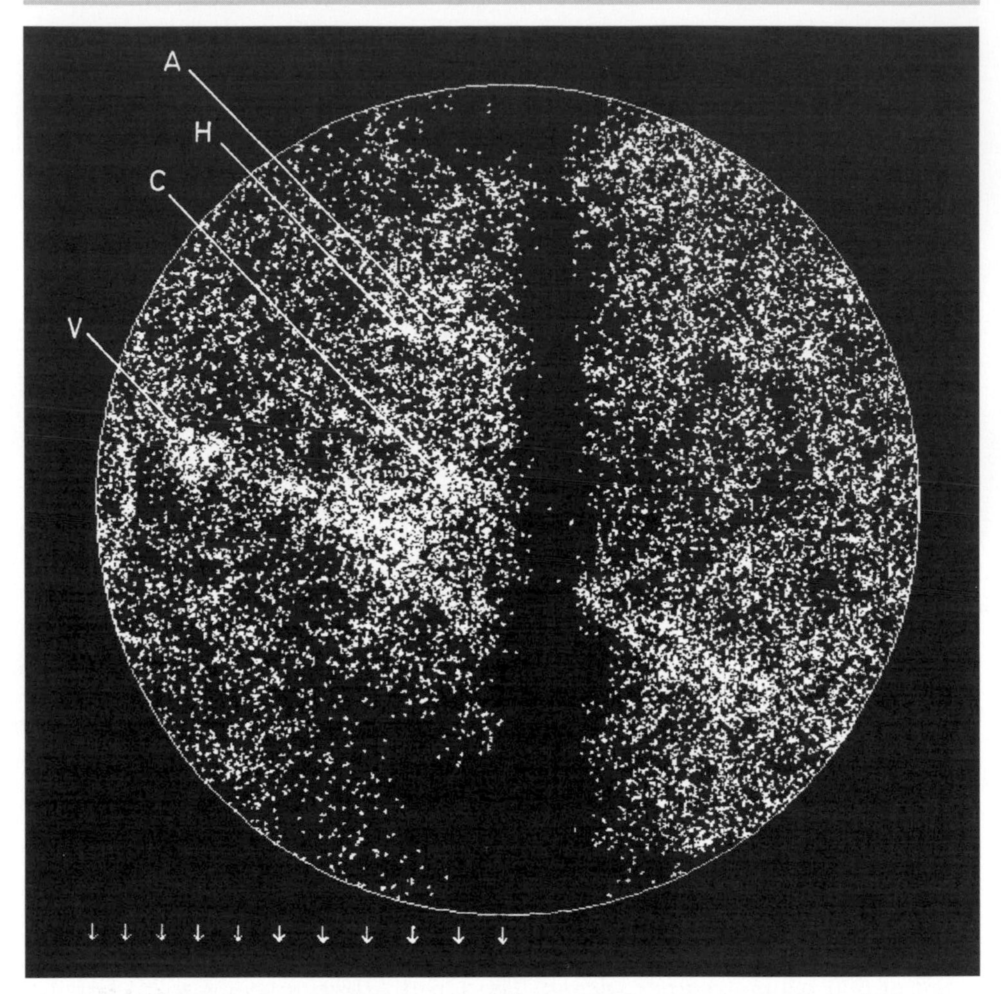

7.1 The distribution of relatively nearby galaxies, in the direction toward which our Galaxy is moving. Various clusters are labelled (A=Antlia, H=Hydra, C=Centaurus, V=Virgo – diagram prepared by Ofer Lahav)

decide whether to abandon Newton's Law on a cosmic scale, or accept that there is considerable dark matter in the Universe.

The consensus for science opted, in the 1980s, for dark matter. There would have to be far more dark matter than luminous matter. Hence, we now find ourselves in the uncomfortable situation of not knowing what most of the content of the Universe really is. Dark matter – mysterious as it is – has come into fashion. The acceptance of dark matter was also driven by the widespread recognition of the *inflationary theory* – the theory (mentioned in Chapter 3) that supposed incredible expansion of the Universe within a split second of its formation. One of its predictions was that the Universe ought to have close to the critical amount of mass required (eventually) to stop its expansion. Yet the visible matter only accounted for less than 1 percent of what was needed. Adding in the apparent dark matter in the outer parts of galaxies and in clusters of galaxies probably took the total to 10 to 20 percent. To make up the rest, some exotic form of dark matter that pervades the Universe on still larger scales was needed. Cosmologists readily debated whether this ought to be "hot dark matter" or "cold dark matter" – a distinction largely in the mass of the individual particles.

Whatever the case, we simply do not know what it is – anymore than we know what the additional mass in galaxies is made up from. There have, of course, been many suggestions, but little worthwhile evidence.

One approach has been to look for *machos*, an acronym for "massive astrophysical compact halo objects" to account for the excess mass in our Galaxy at least. These might be "brown dwarf" stars with very, very low luminosity. Alternatively, they could be "Jupiters" – large planets, but not quite large enough to have achieved stardom. Either way, they possess little mass individually, so the only way they could provide the missing mass would be to exist in very large numbers. In this regard, we are starting to get some evidence from micro-lensing campaigns (explained in the next chapter) and the figures so far are insufficient to account for the missing mass.

A variation on *machos* would be to resort to the perennial astronomical scapegoat – the black hole (again see next chapter for explanation). Abundant mini-black holes throughout the Universe would provide the mass, and have the convenience of not emitting any light. This idea was taken quite seriously in the 1980s, but the failure to detect the predicted radiation from occasional mini-black holes exploding has more or less ruled it out.

The alternative approach concerns *wimps* – for "weakly interacting massive particles". The dark matter might exist in (so far only hypothesized) nuclear particles, that barely interact with the normal particles of regular matter (made of protons, neutrons and so on). At this stage, little can be proved or disproved.

However, in the last several years, the assertion that the Universe ought to have just the critical amount of mass has been undermined. Various determinations (from observations, not theory) now favour a value more like 30 percent of the critical density. This is more easily attainable, but it still requires a major contribution from dark matter.

Or is it that we do not understand Newton's Law of Gravity? Newton devised it to try to solve questions about the Solar System, yet here we are trying to apply it to the cosmos, and still expecting it to work. However, to make stronger gravity, rather than invoke unseen mass, the law would have to be modified. While still working precisely the same at small distances, Newton's Law would have to grow somewhat "stronger" at large distances – at the scale of galaxies, clusters of galaxies and large-scale structures. It would then have to get weaker again or it would disrupt the whole Universe. But physical laws do not work this way, and this suggestion seems too contrived for comfort. The preferred choice, by far, is simply dark matter.

The only possible alternative would be a completely different conceptual approach to gravity to the one used by Newton. As we will see (in the next chapter), Einstein's theory of General Relativity is such an alternative, but it too cannot account for "stronger" gravity on large-scales, except by involving more mass.

Accepting that there is dark matter in the Universe, and that Newton's Law operates as normal, has led to considerable progress in understanding how the large-scale structures came about. Although the time scale is way outside human experience, we can nevertheless simulate their development with computers. Figure 7.2 shows an example. Given the slightest of disturbances in an early Universe – as suggested by that snapshot of the early Universe seen on the inside shell of the Cosmic Egg – gravity works to accentuate and magnify the disturbances by pulling matter in. Gravity makes matter clump, and the clumps become the large-scale structures.

The similarity between the computer simulations and the real thing – the real large-scale structures – strongly suggests we are on the right track. It's all done with gravity, though the question as to where the initial slight disturbances come from is unanswered (and we shall postpone discussion until Chapter 9). However, in this author's opinion, the computer simulations are not perfect. They seem more string-like and lack the bubbly frothy texture seen in the fabric of the real large-scale structures.

If there is gravity operating in the Universe, is there also *anti-gravity?* Only a few years ago, such a question would have been considered scientific heresy, but today it is acceptable. At the time of writing, many are hailing a "revolution in cosmology" – though it seems too early to know whether a revolution has truly happened. Some people consider the evidence tentative, and, as usual in science, spectacular claims are met with a healthy degree of scepticism.

It centres on a type of stellar explosion known as a *supernova.* While most stars – our Sun for instance – will die gently in old age, a few massive stars go out not with a whimper but a bang. They get unstable and blow themselves up. Celestial suicide. We witness these occasional explosions in other galaxies. In 1987, we saw one in the Galaxy next door, in the Large Magellanic Cloud. The intervening dust lanes of the Milky Way hide most of the supernovas that explode in our own Galaxy. Only those relatively close are visible, and history records instances of them being bright enough to be seen in daylight (for days and sometimes weeks). In fact, a local supernova is now long overdue.

A particular sort of supernova (known as Type Ia) is believed to result when

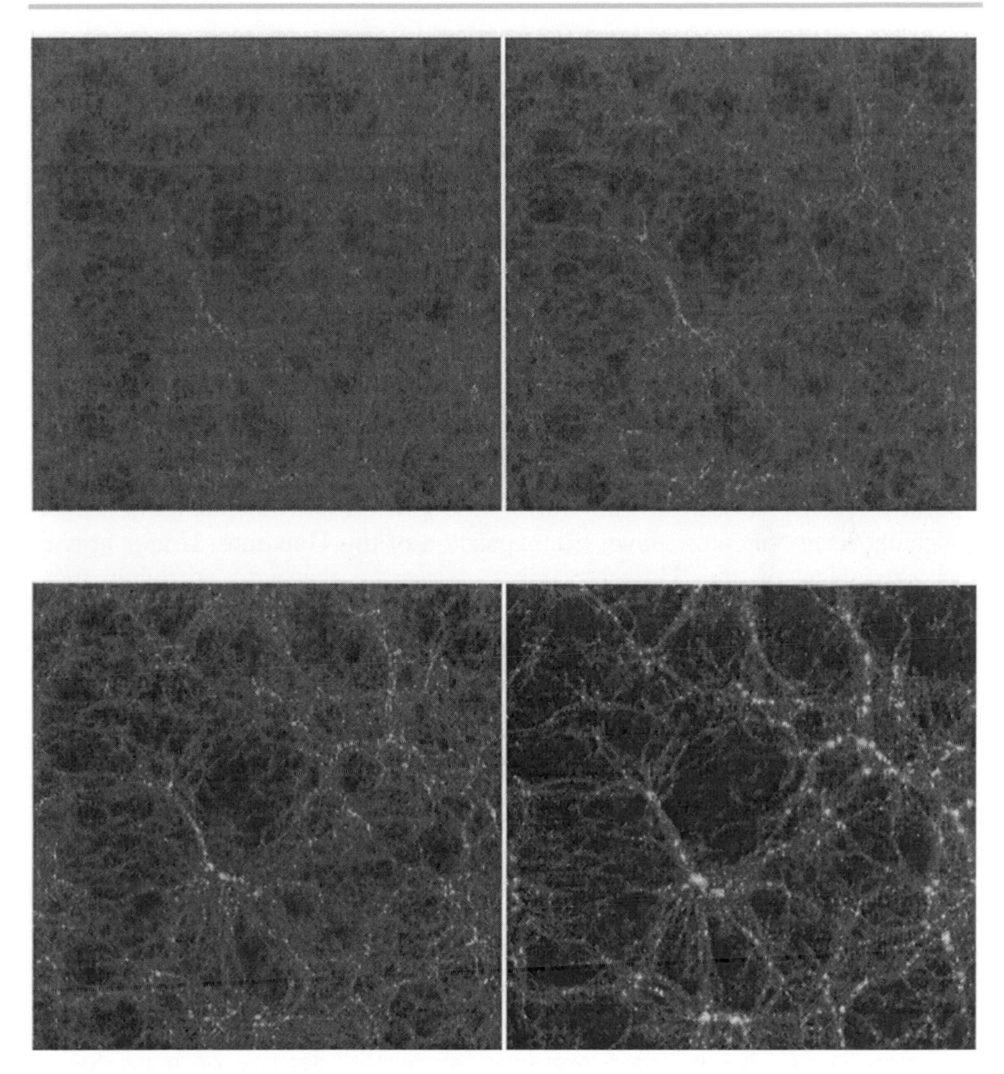

7.2 A computer simulation of the development of large-scale structures in the universe by A. Diaferio and G. Kauffmann.

a companion star tries to dump its material onto a "white dwarf" star. We need not go into technical detail, nor explain how we recognise the particular explosion of the white dwarf that follows. But it is thought to be a sort of standardised scenario – and the explosion is believed to be of a *standard brightness*. In astronomy, something with a standard brightness can be compared with its observed brightness, and thereby used to estimate *distance*. The nice thing about supernovas is that they are so bright, they can be seen even in distant galaxies.

Type Ia supernovas in very distant galaxies have proved fainter than expected. In terms of distance, this suggests that they – and the galaxies that host them – are more distant than expected. The situation would be remedied had the host galaxy a higher velocity of recession (due to the expansion of the Universe) than that we observe. The only explanation is that the expansion of the Universe is not slowing down – but *accelerating*. And that – if the supernova evidence can be confirmed – is a stunning revelation. This is what has been hailed as the revolution in cosmology.

In the next few years, astronomers could show that the distant Type Ia supernova explosions which they observe in distant galaxies are somehow different to those nearby, and the evidence for an accelerating universe could go away. But let us, for the remainder of this chapter, suppose that it does not.

Almost all cosmologists think – used to think – that the only force that operates on a large-scale in the Universe is gravity. Gravity pulls and attracts and must invariably slow down the expansion of the Universe. Under gravity, the expansion of the Universe must *decelerate*; we went through those arguments earlier (in Chapter 3). We questioned then, as cosmologists do – used to do – whether the expansion would be so slowed that the Universe would be brought momentarily to rest, and thereafter collapse.

But accelerating expansion is quite a different matter. It is like gently tossing a stone and seeing it rocket skywards, instead of falling to the ground. It seems bizarre, but it may well be true, that this is the way in which the Universe operates. If so, it means there is something driving the expansion of the Universe. That something is Antigravity.

Unpalatable as it first appears, antigravity may yet prove pleasing in a philosophical sense. It would make gravity more like electric and magnetic forces, which both attract and repulse. The need for antigravity of some sort is also a historical one. Einstein (as we shall see in the next chapter) introduced it.

Some theoreticians have welcomed the discovery. The inflationary theory (Chapter 3) has been under pressure, since the density of mass in the Universe was found to be only some 30% of what it predicted. An antigravity contribution could effectively make up for the missing 70%, and thereby keep the theory alive, though not without complications elsewhere.

Save for Einstein's mathematical speculation, the operation of antigravity is unknown. It is as if Isaac Newton were about to sit under the apple tree again, but this time see the apples fly skyward – this time it leaves us speechless. There is no Newton's Law for antigravity.

Perhaps the key question is whether antigravity is something that pairs with gravity, or is a separate force in itself. Einstein chose it as something separate, a property of the *volume* of space, that stays the same throughout the Universe. Unlike gravity, Einstein's antigravity has nothing to do with normal matter.

The former is strongest where matter concentrates, but the latter is everywhere uniform.

Physicists who study the behavior of nuclear particles seem to be favouring this approach. They suggest that the fleeting appearances and disappearances of "virtual particles" in a vacuum, as predicted by quantum physics, are responsible for antigravity. However, this idea has many objections and the truth is, that if antigravity exists, then no one can yet say what it is, or what it is not. The door is wide open to speculation.

My personal speculation is to ask why antigravity has to be constant throughout space. It seems to me that adopting something that is uniform throughout space is more for mathematical convenience than it is for physical reality.

Again I am drawn to the analogy of the cosmological fruit cake. After all, fruit cakes expand, not because they are given a big bang impetus when they are put into the oven, but because the cook remembered to put in baking powder. Baking powder consists of tiny granules. When heated in the oven, each granule releases gas, and the tiny bubbles so formed create the texture of the cake. It is the growth of the bubbles that makes the cake expand.

Is it the growth of bubbles – the voids in the large-scale structures – that makes the Universe expand? It is almost as if there was a sort of "cosmological baking powder", that created the bubbly texture we see in the large-scale structures. It would explain why the large-scale structures today appear so bubbly and frothy. It would explain what I see as shortcomings in the computer simulations that work with gravity alone. It would explain why so many of the voids in the cosmic texture are approximately spherical. It could provide the driving mechanism for the expansion, and cause the acceleration.

If the expansion of the Universe is accelerating, then the mechanism that causes the expansion is still in action today. The expansion is not entirely due to the big bang beginning. It may be the voids that are growing, and the Universe is behaving like the foam that develops when fruit salts are dropped into a glass of water. It would also explain how large-scale structures form, without the need of large amounts of dark matter. It would work with gravity to create the Universe that we see today.

If this were so, we would still have to understand the voids. Is there anything in their centres? Some of us have looked but not found anything. Or is space itself like a fluid medium, upon which bubbles-like distortions can grow. We shall look at that nature of space in the next chapter.

Personally, I rather hope the supernova evidence for an accelerating expansion does not get shot down. The discovery that there is something so fundamental about the Universe, still to be understood, is like the thrill ... of seeing that apple drop!

Chapter Eight

Einstein's Universe

Newton may have sorted out the Universe sitting under an apple tree, but Einstein did it in the patent office.

Albert Einstein, undoubtedly one of the greatest intellects that has ever lived, recently elevated to Time Magazine's "Man of the Century", nevertheless performed dismally at school. Perhaps that says something of the schooling system, or perhaps something as to how truly great minds develop. In any case, his mediocre record led him to a job as a patent clerk in the Swiss civil service in Berne. It seems the work was not particularly taxing, and left his mind free to work on greater problems. His astonishing breakthroughs led first to the theory known as *Special Relativity*, and later – when promoted to an academic environment – to *General Relativity*. A lesser accomplishment, from his patent office days, was the understanding of the *photoelectric effect*, which earned him the Nobel Prize for Physics in 1921.

While Newton's Law, with gravity making apples fall from trees, seems relatively understandable, Einstein's Relativity is nowhere near as simple. However, relativity, as we shall see, is an essential part of cosmology. It has profound implications on the nature of space and time, and the Universe in general. Difficult though it may be to comprehend, one cannot avoid dealing with relativity in a book on the large-scale nature of the Universe.

General Relativity is an *alternative* theory of gravity to Newton's Law. It ought to have been called Einstein's *Relativistic Theory of Gravitation*. But why do we need it, if Newton's Law works so well? The answer is that under extreme conditions, particularly at velocities approaching the speed of light, Newton's Law and Newton's Law of Motion do not work out precisely correct. Einstein's theory does. Einstein's breakthrough is now seen to be as important a contribution to the understanding of the fundamental nature of the Universe as was Newton's.

But Relativity carries a warning. Many of the predictions of Einstein's theory seem bizarre, or fly in the face of common sense. For instance, finding that clocks may run at different rates for different people, or bending light rays so that they may not be coming from where we think they originate. Undoubtedly, the weirdest of all of these are those that pertain to the "shape" of the Universe as a whole. Things are not what they seem to be. This is the challenge this particular chapter presents.

Let's start by establishing a little common sense, by considering the way you and I look at the Universe. Our world is a *three-dimensional* one, full of three-dimensional objects such as houses, trees and people. We are three-dimensional beings living on a three-dimensional planet.

Dimensions appear to us as fixed and rigid. Once you build a house, giving it a certain width, a certain breadth, a certain height, it stays that way. Short of a serious earthquake, you do not come home one day and find the height of your house has diminished. Bricks and mortar do not do that. The dimensions of space are surely as solid as bricks and mortar. Furthermore, they extend up, down, left, right, forward and back – right out into space. *As we see it,* space seems to extend indefinitely in all directions – so do the three dimensions.

But already, we may have to modify that assessment. We learned (in Chapter 2) that, even on a clear day, *we cannot see forever.* We can only see as far as the shell of the Cosmic Egg. Furthermore (in Chapter 3), we found that the whole Universe is expanding, growing in volume.

The dimensions of spaces are not solid like bricks and mortar, they are like the dough of a cake. They can expand. They could just as easily contract. They can distort and bend. They are elastic. And that's what Einstein's theory is all about!

Unsettling, isn't it? Things are not what they seem to be. The world – the Universe – is not *as it appears.* The last few words are critical. Things appear in a certain way because that's the way we see them. And to see them, takes light rays. It is a fundamental assumption that light rays travel in *straight lines.* Generally they do. Right this moment, I can see a laptop computer in front of me, with a keyboard, and when I reach out toward it, I find I can touch the keys so that this very sentence, which you are now reading, appears on its screen. You the reader can see a book in front of you, and, by touch, you can verify that it is there. Of course, sometimes, light rays can do other things. One thinks immediately of looking into a mirror, or the classic case of the spear fisherman standing in a dugout canoe, having to aim below where he sees the fish, because light rays are bent when they leave the water.

But normally we can depend on light rays. It would be a nightmare if I were to see the laptop computer out in front of me, but by touch find it was off to the side, because somehow the light rays had swept around through 90 degrees.

The Universe is not a nightmare, but we do have to realise that the light rays we receive have not always come in a straight line. In any case, what is a straight line? It is *one-dimensional*, but *what* if that dimension were distorted or bent?

This is the foundation, the revelation, of Einstein's relativity. It's weird because it seems contrary to our everyday world, our everyday experience. It makes us seem less secure. But it opens up a whole new understanding of the Universe.

At the university where I work, General Relativity is taught at post-graduate level. Our students need to have been through various undergraduate courses in mathematics before they can handle the type of mathematics that Einstein's field equations require. I can hardly assume that the readers of this book are similarly qualified. I am therefore going to use an alternative approach to convey, *in diagrammatic form*, the workings and implications of General Relativity. It is not unique to this book, it is well proven elsewhere. Remember that it is a compromise to the much more complicated mathematics.

Let us start with a volume of empty space, as suggested in Figure 8.1. We have shown it as a box. It has width, height and depth. It is, of course, *three dimensional* – even though we have to put the diagrams on to the *two-dimensional* pages of this book.

We need now to convey how dimensions distort according to Einstein's theory. But, in order to make the distortions visible in the diagrams, we are going to have to show only *two* of the dimensions. In Figure 8.2, you can see

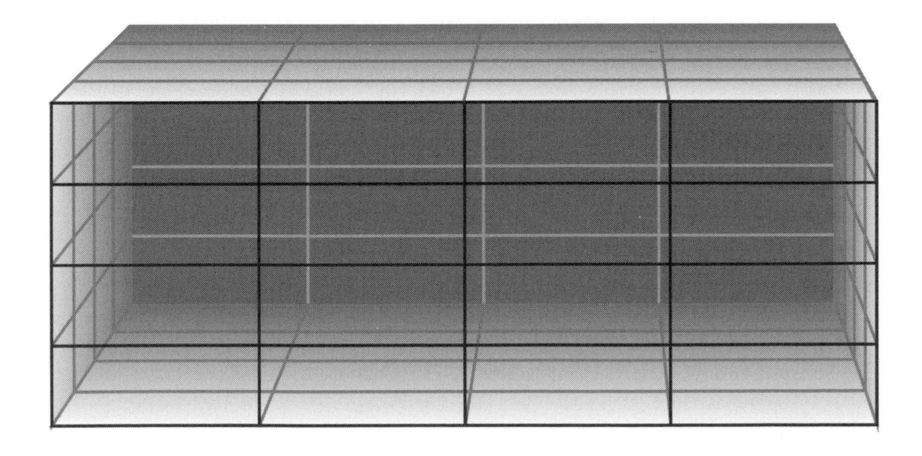

8.1 An empty volume of space

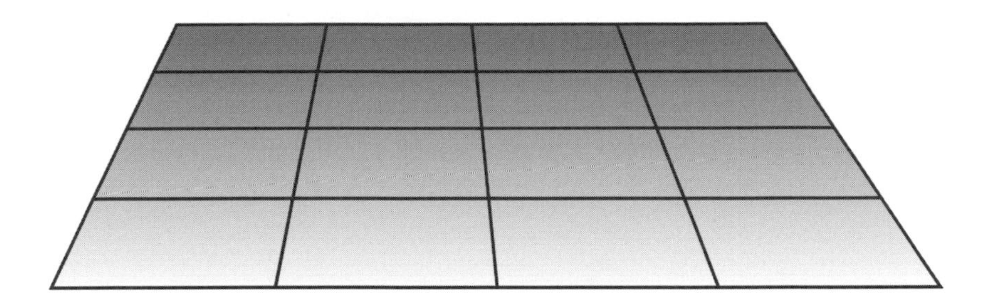

8.2 By suppressing one dimension, a three-dimensional volume appears as a two-dimensional sheet.

8.3 Familiar objects look somewhat strange when reduced to two dimensions.

that we have got rid of *height* and now only have *width* and *depth*. In short, we have suppressed one of the three dimensions. It has been banned from sight, but only as far as the diagrams are concerned. It's really still there. So what you see, in this diagram and the ones that follow, as an *area* is really a *volume*. To reinforce this idea, Figure 8.3 presents a few common objects with the one dimension suppressed.

So far, we have a volume of empty space. Suppose now that we put the Sun in the centre of that volume, then the outcome is shown in Figure 8.4. It's like placing a lead weight on a rubber sheet. The rubber sheet stretches (actually increasing its area very slightly). It deforms. And so does the volume of space that it represents.

Hold on, say the critics! We banned the third dimension just a moment ago, yet here we are stretching things so that they sag – *into the third dimension*.

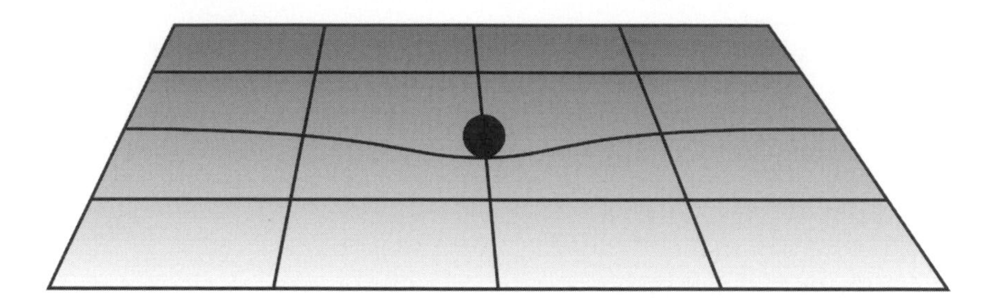

8.4 The presence of a large mass – such as our Sun – causes space to deform

Well . . . the third dimension is still banned, but what we are out to show is that the other two dimensions are *stretched and distorted*. That is what Einstein's theory is all about. Space itself, when considered two-dimensionally, is like a rubber sheet. It is not the rigid bricks and mortar you expected it to be.

Any mass causes the dimensions of space around it to deform. In general, the deformation is very slight. Even with the mass of our Sun, the deformation would be very modest – Figure 8.4 was grossly exaggerated. Your own mass would also deform space, but to an incredibly small extent.

Newton visualised the apple and the Earth pulling each other together by gravitational forces. But Einstein saw any mass as distorting space, and the subsequent motions were the consequence of that deformation. Put a heavy lead weight on a rubber sheet and anything else on the sheet, some glass marbles for instance, will roll toward it. In Figure 8.5, two masses will naturally roll toward one another, not because there is a force pulling them together, but because of the mutual distortion they put on space. This is a perfect analogy of the way in which the diagrams represent the outcome of Einstein's mathematics. So Einstein's alternative view of gravity is to see it, not as a force, *but as a distortion of space.*

It is also a distortion of *time*. We have so far suggested that it is only space

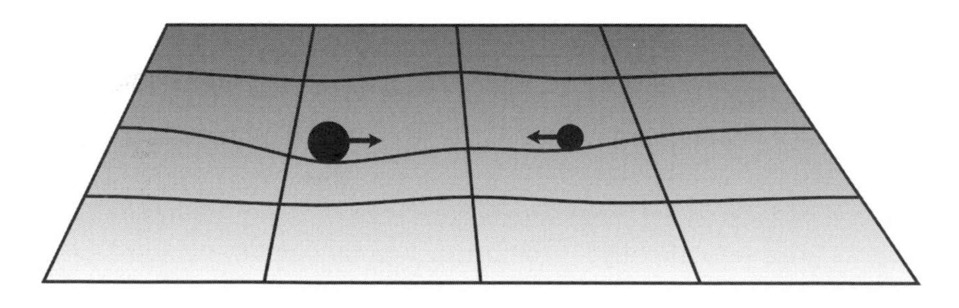

8.5 The deformations caused by two masses cause them to be drawn toward one another.

that gets distorted, but it is *time* as well. Einstein's further revolutionary development was to view time as another dimension. In fact it can be translated into *distance* by being multiplied by the speed of light (since distance = speed × time). Putting a mass into a piece of empty space – and "empty time" – brings about a distortion in space, and a distortion in time.

If you have been able to follow the arguments above, you now understand the essence of General Relativity! Now for cosmology.

The study of *cosmology* is to be able to represent the Universe – its size and behaviour – mathematically. This was not possible before Einstein. After all, in Newton's picture, the three dimensions are rigid and extend indefinitely in every direction without in any way changing. But Einstein allows dimensions to stretch or contract, and that allows the Universe to be "modelled" by mathematical equations.

This is what Einstein did subsequently – but the model he produced suggested that the Universe was not as he believed it to be. His mathematics showed that the Universe *ought to be contracting*! The Universe ought not to be in equilibrium. The stars or galaxies that populated the Universe ought to be in motion relative to one another. Indeed they are. We now know the universe is *expanding* – but Einstein did not. His work was carried out (in 1917) some years *before* the discovery of the recession of the galaxies (as outlined in Chapter 3).

The common understanding at that time was that we lived in a *static* Universe, and the stars or galaxies were thought to be stationary. Yet gravity, in Einstein's universe, was going to cause the dimensions of the whole Universe to shrink. It was going to pull the galaxies ever closer to one another.

The solution to Einstein's problem was (as suggested in the previous chapter) some form of *antigravity*. So he contrived the *Cosmological Constant* – a good euphemism for a fudge factor if ever there was one. The "constant" part of the Cosmological Constant meant that it was the same anywhere, regardless how many galaxies, and how much mass, was present. It simply depended on the volume of space – the very opposite of our speculation. The Cosmological Constant counteracted gravity. It stopped the dimensions of space from contracting.

Einstein was clearly unhappy with his solution, and was to describe it afterwards as the biggest blunder he had ever made. Years later, when Edwin Hubble revealed a Universe in motion, Einstein visited him to learn about the evidence at first hand. Relativity was right, but it wasn't that the dimensions of space were in the process of contracting. Rather something had set the dimensions expanding, and gravity was now trying to slow them down. It was, after all, a universe in motion, not a static Universe. In any case, Einstein was rightfully criticised by the English theoretician Arthur Eddington, for having come up with an unstable situation, since the balance between the Cosmological Constant and gravity was a delicate one: the slightest imbalance would have led to runaway expansion or contraction of the dimensions.

It may offer some comfort to the reader to note that Hubble never quite understood, or at least never quite accepted, Relativity. Hubble saw the galaxies in motion, flying apart from one another. But he did not see the dimensions of the universe, the dough of the cosmological fruitcake, as expanding.

And so Einstein's Cosmological Constant was relegated to the cutting room floors of History. Astronomers thereafter discounted it. But the theoretical cosmologists – those who went on to use Relativity to model the universe – never quite forgot it. It was an option they never quite discarded, and it came booming back in 1998 as the supernova evidence suggested the presence of some form of antigravity.

Einstein's cosmology, tangled or otherwise with the Cosmological Constant, allows us to examine the future evolution of the Universe. The key question has always been whether the expansion would go on forever, or whether the expansion would be arrested and reversed into contraction. The current evidence, particularly if the supernova observations are not explained away, is clearly in favour of a Universe that expands forever. That in itself raises a lot of physical and philosophical issues.

Einstein's cosmology also serves to examine the "shape" of the Universe. The distortion of space by matter imposes a *curvature* to the dimensions. This situation is not unlike the curvature of the Earth's surface. Seen out of a window, the surface of the Earth looks flat. Our senses suggest we are living on a flat surface, but our wider knowledge tells us it is curved. To an astute observer, the curvature is even visible to the eye, when looking out over the ocean. As a ship sails away, its hull disappears leaving only the superstructure visible. Recently, at a beachside cottage, I used binoculars to scan the shipping lane some kilometres out to sea. I saw only passing superstructures, never the hulls.

Einstein's curvature is not unlike the Earth's curvature – except *you cannot see it*. Not only does the curvature bend space, it bends light rays as well. The Universe may well have a curvature, but it would not be visible to those of us inside it, because we could not tell in what way that the light rays had deviated from travelling a straight path.

Three types of curvature are possible – negative, flat and positive. Figure 8.6 shows these curvatures, as they would affect the "two-dimensions-only-please" diagrams used earlier.

Supposedly, the volumes of space we are considering (represented by only two of their dimensions) are very large volumes. On such a very large scale,

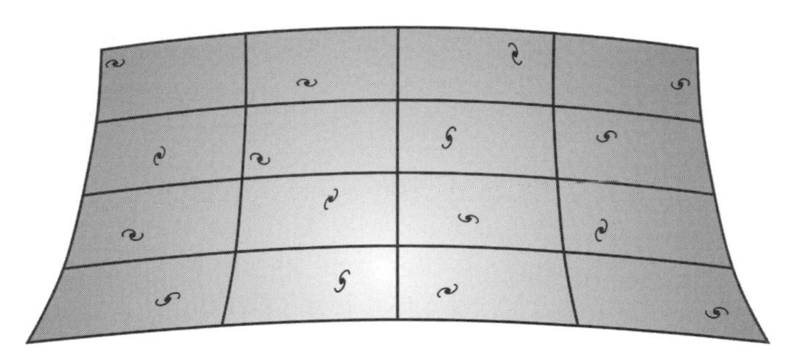

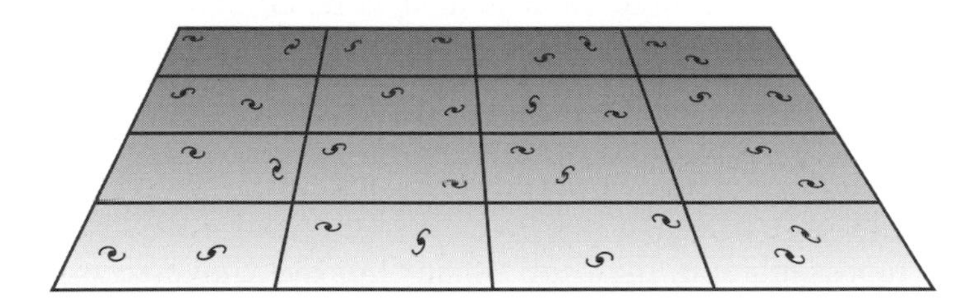

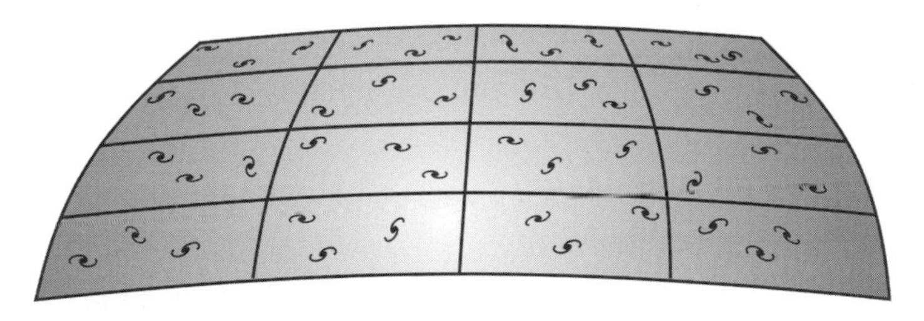

8.6 Scattered galaxies put a gradual curvature on space. As their density increases, so the curvature progresses from negative to flat to positive.

even the large-scale structures repeat often enough that the distribution of galaxies can be considered as uniform. One can imagine the galaxies spread like *hundreds and thousands* on the surface of an iced birthday cake. Two factors influence the curvature of that surface – the spread of galaxies, and ... antigravity.

If, for the moment, we discount antigravity, then a light sprinkling of galaxies will bring about *negative* curvature. A heavier sprinkling would cause the curvature to progress from negative, to *positive* (as was suggested in Figure 8.6). Ever since Einstein's revelation of the bending of the dimensions, the main quest of cosmology has been to discover whether the Universe has negative, flat or positive curvature. The crucial question is how heavily the galaxies are sprinkled. The dividing line between negative and positive curvature is the *critical density* – the same as we spoke about earlier (in Chapters 3 and 7), when we reported that it seemed the Universe had only 30 percent of this value.

The inflationary theory – which supposed that the expansion of the Universe started with a initial great rush – predicted that the density of the Universe should be exactly equal to its critical value. In recent years the theory had run into hot water as the measured values failed to reach that amount. However, theorists will again be happy if both the contribution of matter and the effective contribution of antigravity, together add up to 100 percent (as mentioned in the previous chapter).

If the combined contribution is close to, but does not quite reach 100 percent, the Universe has mild negative curvature, and if we extend our two-dimensional representation, then the "shape" of the Universe would be that shown in Figure 8.7. In that figure, we have to put a boundary to the warped sheet; it comes to an edge, but this is only because we have limited space in which to draw the figure. In reality, it would not come to an end. The Universe, by Einstein's measure, extends forever – an infinite volume.

Trying to conceive an infinite Universe causes indigestion. Is the whole thing populated with galaxies? An infinite number of galaxies? An infinite

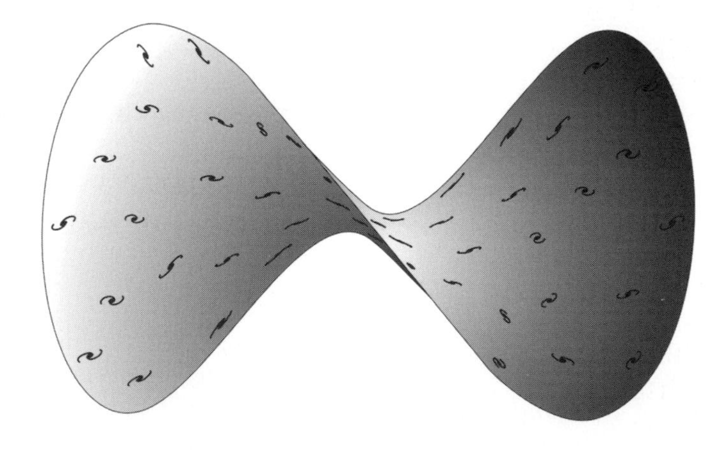

8.7 A universe with negative curvature. Although the diagram suggests a boundary, there ought rather to be no "edge" to the universe.

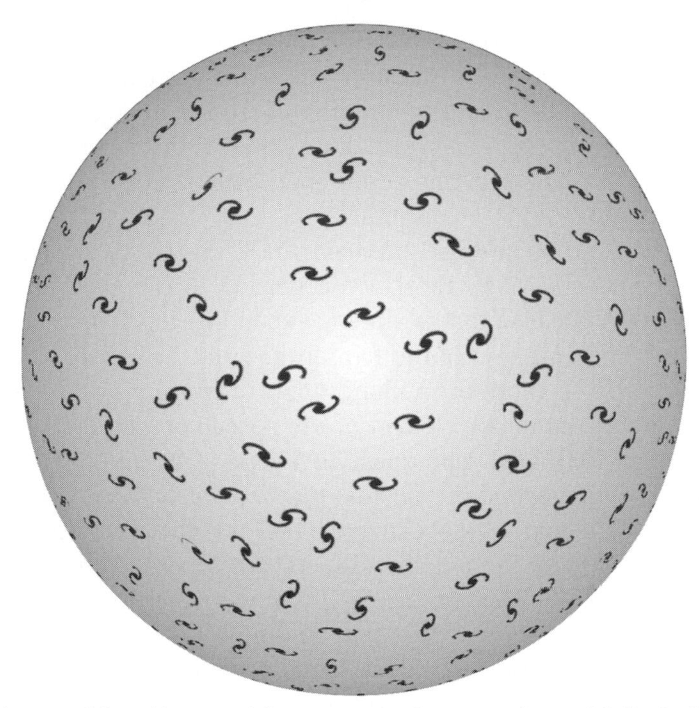

8.8 A universe with uniform positive curvature forms a sphere with limited sufrace area, representing limited volume.

number of stars? Where do we fit in? Fortunately, we can take comfort in that the Universe we see is far from infinite. It is contained within the Cosmic Egg. Does it really matter whether the Universe outside the shell of the Egg is infinite or not?

No doubt time will tell if we have got our measurcments (as to how densely mass is sprinkled in the Universe) right or wrong. If the present measurements are wrong, and if antigravity, if it exists, is not too strong, then perhaps the Universe has mild positive curvature instead.

If the combined contributions from matter and antigravity just exceed the "100 percent" level, the Universe would instead have mild positive curvature. Positive curvature makes a big difference. If the curvature is everywhere uniform, then the two-dimensional representation can be extended, such that it closes into a sphere, as shown in Figure 8.8. It then closely follows our analogy to the Earth, the surface of which also has positive curvature. Space literally meets up with itself. The surface area of our sphere is finite, not infinite; so too would be the volume of the Universe.

Moreover, it would be just one dimension more than our experience on Earth. While the Earth that we stand on looks flat, we know we can sail off in any direction, go right around the Earth, and come back from the opposite direction. So too with a positively curved universe. Jump in a spacecraft and head off in any direction, and if the universe were not expanding, you should eventually return from the opposite direction. Like travelling on the Earth, you would not be aware of the curvature. You would think you were going in a straight line, but the fact that you would eventually return to the same point, would be the proof.

But such proof is not possible because the Universe is expanding, much faster than we could possibly travel. Even getting to the shell of the Cosmic Egg is not possible, because that is already expanding too fast – and that might have been only the beginning of the trip around the Universe.

Einstein's theory is therefore talking about curvature of the dimensions, brought about by the cumulative effect of the spread of galaxies, *on a very very large scale*. It has long been our quest to measure it, and thereby to know whether the Universe beyond the Cosmic Egg was likely to be finite or infinite.

It is also possible that, if the Universe had positive curvature, and if the effects of antigravity were limited, then the expansion of the Universe might be gradually arrested, and reversed. The expanding Universe would switch to a contracting Universe. If so, the Universe, in the way we know it, would have a finite lifetime. After starting from a big bang, it would eventually end in a big crunch!

However, the present evidence is that the Universe has insufficient mass and sufficient antigravity that it will never re-collapse. The lifetime of the Universe appears infinite. The nature of the evidence, and the nature of science, is that future measurements could indicate otherwise, but for now that seems unlikely. The picture of the Universe that it presents raises all sorts of philosophical questions.

We may not be able to see how relativity has affected the shape of our Universe, but we can witness the effects of relativity on somewhat smaller scales. Figure 8.9 shows a cluster of galaxies – a concentration of mass strong enough to distort the space around it. Light rays that have to pass through that distorted space, such as those from more distant galaxies, are bent. This has resulted in a mirage effect, with distorted multiple images of the distant galaxies.

The same effect can occur with the stars in our Galaxy. The deviations in position are much smaller, and cannot be recorded by a telescope. However, an increase in brightness of a star is observed if another star or foreground object passes directly across our line of sight. One product of the surveys that look for this "micro-lensing" is that they have not found enough of the dark matter "machos" (Chapter 7) as they should in this way.

Finally, let us look at Relativity on a very local scale. Soon after Einstein published his theory, the mathematician Karl Schwarzschild saw another

8.9 A demonstration of relativity: the distorted images of distant galaxies have been caused by gravitational lensing (Hubble Space Telescope).

consequence. If space deforms as matter is put into it, putting in ever greater densities of matter ought to make it deform even more – until it reaches its limit!

An analogy is a simpler way of putting it. Place a heavy weight on an elastic sheet, and the sheet will sag in the fashion suggested earlier (Figure 8.4 for example). Put more lead weights, and the sheet will be stretched almost to the limit. Still more weights, and it will start to stretch uncontrollably, in the process forming a hole!

Schwarzschild was the first to predict the existence of *black holes,* as a consequence of Einstein's Relativity. If we retain our two-dimensional visualisation, then the outcome is the sort of geometry depicted in Figure 8.10. It is called a black hole, because anything that has the misfortune to roll in will not normally ever come out again – and that includes light. Even light cannot escape, and so the "hole", created by overloading the elastic dimensions of space, appears *black.*

Of course, it sounds almost crazy to predict something that cannot be seen! It's like saying a hall is crowded with invisible men. How can you prove they are there? Well, black holes may not be seen – but anything that ventured close to one would nevertheless feel its strong gravitational pull. Like celestial vacuum cleaners, they would try to suck in everything around them.

The idea that black holes might exist was given a strong boost in the late 1960s when *neutron stars* were discovered. They were unbelievably dense. A teaspoon of their material would weigh several million tons! They showed that such dense concentrations of mass were indeed possible in the Universe. Moreover, they were very close to becoming black holes. Because of their

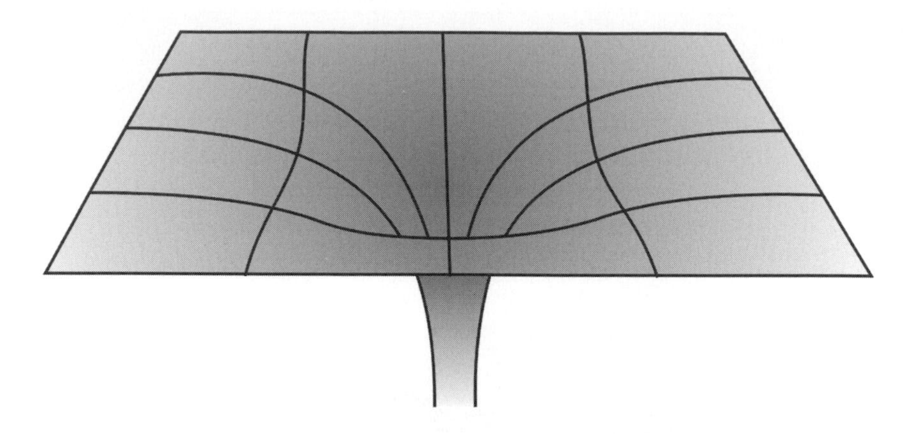

8.10 A black hole. Space is so distorted by a dense mass that it seems stretched into a tube.

extreme gravity, neutron stars have subsequently proved to be good testing grounds for Relativity – and results are as Einstein's theory predicted.

Black holes were predicted to occur in two astronomical situations. The first – like the neutron stars – as the remnants of old massive stars that had collapsed when their energy sources were exhausted. Indeed, we know of a handful of binary stars – orbiting around each other – where one component is massive enough, and invisible enough, to be a black hole. Even so, astronomers have been disappointed not to find such systems in greater numbers.

The second situation, which has proved to be more likely, is in the centres of galaxies. Here we are talking about truly massive black holes, millions of times more massive than stellar remnants. They were first predicted as a way of explaining the properties of *quasars* (encountered earlier in Chapter 5) which often have extremely energetic nuclei. Enormous energy is released from matter falling toward a black hole. Thanks particularly to the Hubble Space Telescope, we now have many cases where stars extremely close to the centres of galaxies are known to be moving at very high speeds, apparently in orbit around a central, yet invisible, mass. Our own Galaxy shows something similar.

So black holes, bizarre as they seem, almost certainly exist. In them, the elastic dimensions of space have been stretched to a tube. But, if space itself cannot come to an end, where does that tube lead? We do not know, but speculation is rife. Einstein himself, together with his collaborator Rosen, worked on the problem in the 1930s. His mathematical approach suggested two solutions implying that the tube should lead to another funnel-shaped opening in another flat sheet – as suggested in Figure 8.11. But where was the

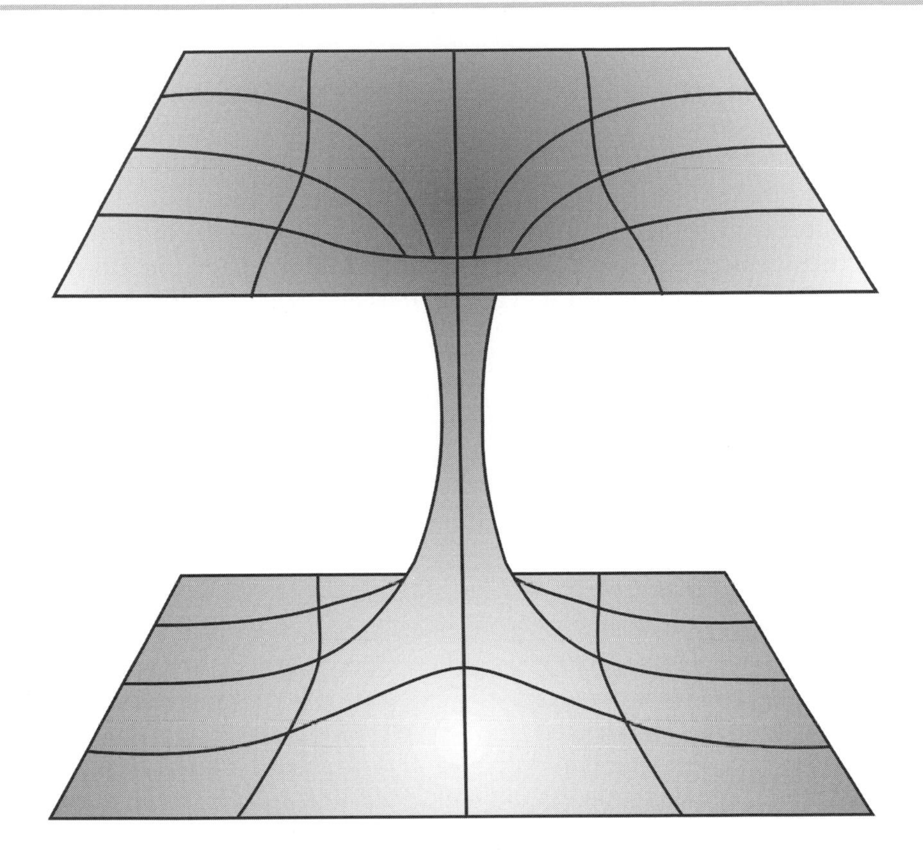

8.11 A wormhole. Two black holes may link two different points in our Universe, or even two universes.

other flattened sheet? Could it be somewhere else in our Universe, or could it even be in another universe?

The Einstein-Rosen "bridge" suggested a connection from one part of the Universe to another. Of course, it is complete speculation, but science fiction writers have taken up the idea ever since. It offers a convenient way to jump from one place to another. When the spaceships of the "Evil Empire" are about to catch you, head into distorted "Hyper-space" and pop out somewhere else, completely safe.

But is it as easy as it appears in science fiction? No, any foolhardy astronaut who fell feet first into a black hole would have his feet accelerated much faster than his head by the extreme gravity. He would be "stretched apart", and then likely crushed to an infinite density! Reality is not as smooth as science fiction.

However, you might just avoid being crushed if the black hole is rotating! Mathematical solutions, first derived in the 1960s, indicated that there might just be a way for something to fall through a black hole and emerge on the other side. Again science fiction rushed in, and the finding quickly became the concluding sequence of Stanley Kubrick's *2001 – A Space Odyssey*. Conveniently, science fictional characters seem unaffected by the extreme gravitational forces that make the real journey impossible.

But, short of jumping into a black hole, we still need to know what happens to that tube of distorted space. Does it lead to another part of our Universe, or does it even interconnect two universes? Perhaps *Science will never know.*

Chapter Nine

The Limitations of Science

It's an incredible situation. Here we are, confined to a tiny dot in the cosmos. In an observable universe of some *15 billion light years* radius, we talk of "going into space" when the Space Shuttle travels *a thousandth of a light second* from the surface of the Earth! The furthest man has gone is about one light second, and his robotic spacecraft some 10 light hours. Yet you might well get the impression *we know all about it*!

Far from it. There is a considerable amount that *we do not know* about the Universe and the way in which it works. Worse still, there must be things we do not know, which we do not even know that we do not even know, and therefore cannot discuss!

So far, in this book, we have generally dealt with what we do understand about the Universe on a large scale. It is equally important to understand how much we do *not* understand. For example, what is the eventual fate of the Universe? What lies beyond the Cosmic Egg?

It is nevertheless impressive how much we *have* found out about the Universe around us, simply from looking out from the surface of the Earth. It is astonishing how much has been learned in less than a hundred years. That we know roughly the nature of the cosmos, that our knowledge has penetrated right out to the shell of the Cosmic Egg, are incredible achievements.

It is also impressive how much we can understand about places we cannot see – like the interior of the Earth, or the interior of the Sun. There our knowledge of the laws of physics, originating from laboratory work, has let us predict temperatures, pressures and the like. It has allowed us, for example, to understand how stars work – how the extreme temperatures and densities in their core permit nuclear fusion. Hydrogen fusing into helium provides the energy that keeps almost all stars shining; it keeps the Sun shining, and the flowers growing! Such has been the progress of knowledge, such has been the progress of science.

Science works by observation, hypothesis, test and acceptance. Newton's Law and Einstein's Relativity are excellent examples of how the system operates. New ideas, such as both Newton and Einstein had, may at first seem controversial. Before gaining general acceptance, they have to be thoroughly tested and conclusively shown to work. Many theories are disproved and discarded. Science rests on proof, not belief.

The past century has been a scientific age. Progress has been so fast, that it is tempting to believe that it will not be long before science understands everything about how the Universe operates. But there is one limitation, for science cannot measure how much more there is to learn. Overconfidence has, in the past, led us to believe we were close to knowing everything, when that was far from the case. In Victorian England, there was a feeling that there was not much that still needed to be discovered, and that there was really no future, since (as expressed by the author Galsworthy) *England then expected the present to endure*. Whatever then would Victorians have thought of radio, television, jet planes and the Internet? Even in the past twenty years, Stephen Hawking (one of the world's most celebrated scientists) speculated that we were close to knowing everything about the laws of physics. But there are still many unresolved problems in particle physics, which we have to overcome – and those are only the problems we know about.

It is human nature, however, for humans to *think* they know about everything – the unknown engenders insecurity. Many individuals gain self-confidence, and status within society, by projecting an image that "they know it all"! These are the people never short of answers, sometimes pleasant, sometimes somewhat arrogant. You meet plenty in everyday life. Quite obviously, the scientific community has its share of these sorts of individuals. They may sometimes convey the impression that science knows it all – when that is far from the case.

The general public doesn't understand scientists, and scientists generally do not understand the public. All too often the image of a scientist to the public is either that of a backroom boffin, whose mind is so occupied and his hair uncombed, that he has to be told when to eat – or a somewhat overconfident "I know better than you" personality, who has deigned to descend briefly from his ivy tower to speak to the mere mortals.

Perhaps that is exaggeration, but it does reflect a general and global problem, as to how the general public perceives scientists. Of course, years ago, scientists were also helping to engineer nuclear bombs or building rockets, so politicians, at least, had to listen. Happily, the Cold War is over, but it has meant that scientists have lost a lot of political clout. Many modern politicians do not see the need for science. Certainly, the standards of scientific education in the United States have slipped alarmingly. Many Americans now see scientists as almost a religious sect, with its own program and ambitions.

This author's feeling is that scientists ought to communicate more with the public (hence this book) and be more honest, freely admitting what science does not know (hence this chapter). Openness is particularly necessary as cosmology opens up far more questions than we can answer. Let's start by reviewing one of the major problems.

Science has spent a couple of decades hitting its head against a brick wall – except we have been unable to see that brick wall. I am, of course, referring to *dark matter* – as already raised in the two previous chapters. It is perhaps the most disturbing element in our knowledge of the Universe on a large-scale, because it ought to be something we could solve. But either it is there, and we do not understand what it is, or it is not there, and we do not understand gravity.

Yet gravity is something science has confidence that it does understand. We have seen the remarkable success and enlightenment that Newton's Law of Gravitation brought to our understanding of the cosmos. Here was something that could be tested on a very small scale, even within a laboratory, yet applied to the planets in the Solar System, double stars and to ever-larger scales. Einstein's Relativity, though conceptually different to Newton's Law, was a refinement, to take into account the slight distortion of the dimensions of space. Einstein is correct to the best accuracy we can measure. No theory has been able to get the better of General Relativity, and none needs to. No further refinements appear necessary.

Hence the alternative, hence dark matter. It has led science to speculate on the existence of *machos* and *wimps,* the idea that the dark matter might be held by zillions of Jupiter-like planets, or reside in massive nuclear particles. This thinking has been around now for some twenty years, yet the only progress we can speak of is in what the dark matter may not be, rather than what the dark matter actually is. We are, very literally, groping in the dark and this is a major limitation to our understanding.

A few scientists, less conventional than the rest of us, mock the idea that we have to invent a kind of unseen mass. They claim that we ought to realise that our whole picture of the Universe may have to be revised – either the laws of gravity are wrong, or the general scenario of the expanding universe is incorrect. They even suggest that the Universe is not expanding, but that light rays tire, lose energy and give the impression that the Universe expands. These, however, are radical ideas.

It all comes down to trying to make the best judgement on the evidence presented. No one can say what is absolutely right or absolutely wrong. Scientists like myself would rather say there seems, on the basis of probability, a better than 99 percent chance that the expansion of the Universe from a big bang beginning is real – and that's where we would place our money.

The nature of the extra mass is the biggest *current* problem cosmology faces. I wonder what I might be able to say in thirty years time? It's funny the way things change, because thirty years ago researchers would have said the biggest problem was something else.

I am referring to the eventual fate of the Universe. Will it expand forever, or will its expansion be stopped and reversed into a contraction? At least we have made some progress in answering that one, though our conclusions are nowhere near the 100 percent level of confidence. As already outlined, estimates of the mass present, even with the extra mass included, still put it as insufficient to provide enough braking power to stop the Universe. If antigravity really exists, then the Universe is even less likely to stop. This means the Universe expands forever.

However, there is still a chance that antigravity may add to the mass, enough to give the Universe positive curvature, so that it might be proved *finite* in extent (as was shown in Figure 8.8) – a bit like finding the Earth was round and no longer flat. Nevertheless, the Cosmic Egg shows that there is a limit to how far we can see, so in a sense whether the Universe beyond its shell is finite or infinite now seems less important. Also, the Inflationary Theory, favoured by many cosmologists, predicts that the bending of space is very very mild, so the Universe is vast anyway, and the Cosmic Egg encloses only a very small part.

The problem hasn't gone away, but its priority has slipped. Cosmology is about what we do not know, as much as what we do know. But from the problems we ought to be able to solve, let us now move to the problems that we may never be able to solve.

Is the portion of the Universe that we see just like the rest of our Universe? We would like that to be the case, but if the Universe is infinite, then what we see – the entire contents of the Cosmic Egg – is an infinitely small portion of the Universe. I have jokingly suggested that the world beyond the Cosmic Egg is full of liquorice allsorts – because no one could ever disprove me! We will never see beyond that Cosmic Egg, so we will never find what is out there.

So we have only a limited view of our Universe. It is like standing at only one spot on the Earth and hoping to describe the entire planet. If we stood in the Sahara, we might think the whole planet was desert. If we stood in Antarctica, we might think an ice sheet covered the Earth. Of course, the chances are that we would see ocean, and think the whole planet was covered in water, which at least is 70 percent correct.

But we cannot say, with any certainty, what we might expect to find beyond the shell of the Cosmic Egg. Because the Universe that we see is full of galaxies, our best guess is that the Universe continues in the same fashion – zillions and zillions of galaxies we will never see, and never know.

Space seems infinite in extent. Is time the same? So entrenched is the perception of time to our minds, that we also see it stretching to infinity, both to the past and the future. Tell anyone that there was a big bang beginning to our Universe, and they will often ask "Ah yes, but what was there *before* the big bang?"

Possibly nothing, not even space, *not even time*. Did the clock of time only start running with the big bang – or was it running already? Einstein's approach gives us good reason to believe that time is a property within this Universe. If so, time only began when this Universe began. It didn't exist before the Universe. The big bang is not just a big explosion. It is the moment of birth of the dimensions of space. It is *possibly* the moment of birth of time.

Nevertheless Einstein's theory gives us a way of understanding the basic dimensions of our Universe, both space and time. It tells us that space and time are part and parcel of our Universe – not that our Universe itself *has to dwell in space and time*. Effectively, all of space, all of time, is confined within our Universe. That, after all, is what a Universe is.

The theory could, however, equally well apply to *another* universe – a completely separate universe to ours – with its own system of space and time. Would space and time behave the same in the other universe? Would there still be three dimensions of space and one of time – or more, or less? Could there still be another universe, and another?

There could be a multitude of different universes – each a world with its own space and own time. Our Universe may only be one of many. It's a daunting prospect, but one we need to consider, because while there is no way to prove it, there is no way to disprove it.

Philosophically, the idea has been around for some centuries. But the old idea was one of *island universes* – repetitions of starry systems like our own. One possible interpretation of the nebulae found scattered across the sky by William and John Herschel was that they could be such island universes, too distant for individual stars to be seen. Indeed, this was so – most of the nebulae eventually proved to be external galaxies, thereby proving the hypothesis correct.

But the sort of other universes we suggested above cannot be seen. They do not exist within the same system of space and time system as we do. How do you know that something exists if you cannot see it? You do not. But you also do not know if it doesn't exist.

Einstein's Relativity did, however, lead to some intriguing speculation. A rotating black hole has so contorted the normal dimensions of space and time, that (as described in the precious chapter) it entangles a labyrinth of tubes of space and time. The speculation has it that these "wormholes" connect, either to other points in our Universe, or to points in other universes. But such wormholes are merely speculation, and not in any way proof. Even if they did exist, travel through them would still be far too violent for man or machine.

The question concerning the multiplicity of universes is one of the most profound. Yet we see no way of answering it. It is, in a way, the greatest limitation of science.

So far, in this discussion, we have dwelt with the Universe at its largest scale. Yet we could equally well question things *on a small scale*, as we did on the large. Though the former is generally outside the emphasis of this book, we must nevertheless inter-link the two. Everything that today is very large was, in that first fleeting moment of the big bang, very small. Today, our understanding of conditions within split seconds of the big bang itself comes not from astronomers but from particle physicists – those who examine physics on a very small scale.

Whereas we astronomers work with planets orbiting stars, and stars conglomerating and swirling around in galaxies, these physicists rather see nuclear particles conglomerating into atomic nuclei and atoms. The interplay of mass on this scale is far removed from Newton's gravitation. Rather, electromagnetic and nuclear forces rule.

Einstein's Relativity and the nature of the Universe seemed somewhat bizarre compared to everyday experience and everyday common sense. But Nature is even more bizarre when we turn to how it operates on a very very small scale.

After all, what is matter itself? We think of matter as somehow solid – atoms packed tightly together. Yet, if we took an atom to be the size of a hot air balloon, we would be astonished to realise that it was almost all empty space. Science tells us that almost all (99.9 percent) of the mass in an atom is packed into its nucleus, and the nucleus in our analogy is no larger than a grain of dust! The rest is just hot air. Not really, the rest of an atom is an electron cloud – very very tenuous. Given the incredibly small size of the nucleus, the mass inside it is packed at incredible densities.

At this level, physics has identified particles. Electrons, protons and neutrons are the more familiar, but there are many more. Further, protons and neutrons are themselves made of particles called quarks. Yet the word particle suggests a small solid speck, and that is misleading. We have found that these particles also behave as waves – each a little packet of energy.

Wave mechanics has allowed us to understand how these particles operate – and to predict successfully how matter works at this nuclear level. It has also been able to offer considerable insight into the behavior of the Universe in those first seconds after the big bang beginning. For instance, it allowed us to predict by calculation that within the first three minutes, all the matter would be turned into 77 percent Hydrogen, 23 percent Helium, and very very little else. That prediction appears to be correct!

Wave mechanics – also known as quantum mechanics – does not operate according to fixed laws, but by ***probability***. The stronger the wave, the more likely the particle is to be there. It seems more like "Alice in Wonderland" than reality, but it is reality on a very small scale. One can at least take comfort that even Einstein did not like it. His famous remark "God does not throw dice" expresses his view, that neither he nor his Creator like to gamble, to live by chance or probability.

So here we have another limit. How do we reconcile Einstein's Relativity, which makes precise predictions as to how things move in the Universe, with quantum mechanics, which provides only probability as to what is going to happen? For years, particle physicists have talked about a *Grand Unified Theory* that would bring even gravity into the realm of particle physics. But it is still a hypothesis, because to test it, we would need nuclear accelerators far more powerful, and far more expensive, than those we have at present.

It is this grand unification that led to the Inflationary Theory (described earlier in Chapter 3). It predicts that the early Universe inflated enormously in size in a split second! It is an attractive theory that explains the character and uniformity of the cosmos today. But we have no direct way of testing whether it really happened.

Yet, we have barely scratched the surface on the limitations of science. So far, we have only been concerned with physical science – the science that shapes the cosmos. Yet *life science* is far more complicated. Bacteria, beetles and buffaloes have far more complex internal structures than stars! Moreover, although we have made enormous strides in understanding how life operates – DNA codes for example – we still do not know quite what life is and how it began. We do not know what consciousness is, on a scientific basis.

The mention of the *origin of life*, for many, brings in the role of a Creator. For the questions that science cannot answer, people have usually turned to religion. The mainstream religions of the world – Christianity, Islam, and Buddhism – offer scenarios as to how the Universe began, to how life began. Many sincere individuals have tried to reconcile the scientific findings about the cosmos with the religious accounts.

Sometimes it proved dangerous ground. One of the great intellectual conflicts of history was when science (through Copernicus, Kepler and especially Galileo) first proved that the Earth and other planets orbited the Sun, when the Church of Rome held that all else went around the Earth. In retrospect, the Church had taken too literal a translation of certain statements in the Bible, but at that time, it was heresy to think otherwise.

It is still dangerous ground. Fundamentalist Christians in the United States have for years tried to stop the teaching of the Theory of Evolution – the basis for the scientific understanding of the diversity of life on the planet. In their picture, the Earth was created "as is" only a few thousand years ago. Though they are sincere people, science finds their scenario impossible to accept. If the fundamentalists were correct, it would present an image of the Creator as a deceiver – who, for instance, seeded the Earth with the dummy fossils of other animals, which never existed.

It would also have cosmological implications. Were the Universe no older than a few thousand years, then we ought only to be able to look no more than a few thousand light years into space. That is not far enough to even see the Galaxy we inhabit, let alone the billions of other galaxies – unless the Universe was created *with the light rays already in place*! There would then be no need to create the galaxies, only the light rays – a great saving of trouble for the Creator. But on such a basis, the Universe could have been created *just ten minutes ago* – all the light rays in place, and your brains preprogrammed with memories and all!

Clearly science and fundamentalist Christians clash. But the latter are a minority. The bulk of the Christian community, and most other religions, work in harmony with science. Religion is, for many, a way of exceeding the limitations of science. It gives a purpose for existence and it explains why the Universe is the way it is. It needs to be seen alongside the scientific picture.

Although many of the world's religions are very different, there are many common traits, especially in regard to the cosmos. Most religions and science now agree on one cosmological point, which is that the Universe had a beginning. We have already presented that argument scientifically (in Chapter 3). The development of cosmology in the 20th century has altered the old view that the Universe was static and unchanging in time, to a Universe that is evolving – ever since its hot big bang beginning.

Many people have tried to reconcile the religious and scientific accounts of the beginning of the Universe. One author, at least, has suggested that Einstein's Relativity could be used to slow time, to make evolution match the Biblical seven "days" of creation. However, the compression of billions of years into mere days requires something to travel at a speed barely less than light itself – and the "days" of the early Earth were very much shorter than they are at present.

Given the limitations of science, many individuals see religious beliefs as complementary – filling in those aspects science cannot address. There is a great deal of harmony, more than there is conflict. In the last chapter we move, not to a religious discussion founded on belief, but to a philosophical discussion based on evidence. In the concluding pages, we will explore where and when we exist. Against the large scale of the Universe, we seek a perspective of where we fit in. In doing so, we may also gain a little understanding of the meaning of our existence.

Chapter Ten

The Anthropic Principle

In traditional legend, the Zulu people saw the sky as a blue rock, which covered the Earth. Upon the face of that rock moved the Sun by day and the stars by night. It seems delightfully simple, but it is the perception of the Universe that most humans have had – a flat land below, and a domed roof above – a small and comfortable world. It is the perception that most other creatures of the planet must have.

But as humans began to travel, so they became aware of the *size* and *curvature* of the Earth. Though Eratosthenes (the head of the great library in Alexandria) had accurately determined the circumference of the Earth in 200 BC, only 500 years ago the estimates had drifted wildly off. Enough that Christopher Columbus sailed confidently off into the blue, expecting to find the far east, only a few weeks sailing to the west of Europe. Modern maps that show both "East Indies" and "West Indies" – which later proved not to be the same – are testimony to this folly. Only when the survivors of Ferdinand Magellan's expedition returned to Europe in 1522 – the first persons to circumnavigate the Earth – was the size of the planet properly appreciated.

The world had grown larger, but not uncomfortably large. It still remained the dominant central body in man's Universe. But the age of discovery in Europe brought about both exploration of the Earth and exploration of the ways of nature. The Europeans were first to make a scientific assessment as to how the planets in the sky moved, and the only sensible and correct interpretation was that the Earth orbited around the Sun with the other planets. It was a stunning relegation of the Earth, and (as already mentioned in the previous chapter) was not lightly accepted by the political powers of the time, for man had always held central place in the Universe.

Since then, even the Sun has lost its position as the central body of the known Universe. The realisation that the stars were suns, and our Sun simply one of them, crept in slowly and scientifically. The lack of fuss, in this regard, is a reflection on public understanding; the news has yet to reach everybody even today! Suffice to say, it was no longer seen as undermining any religious or political authority.

With telescopes, the Sun was downgraded to being just one amongst millions of stars; but even at the dawn of the 20th century, most astronomers believed it nevertheless held a central place in a Universe of stars. However, that was no more than claiming we were in the centre of a wood because we could see trees all around. We could not see the wood for the trees, and we could not see the Galaxy for the dust. The dust clouds in the Galaxy restrict our vision when we look "horizontally". Only early in this century (thanks mainly to the work of Harlow Shapley), did we discover that we were far removed from the centre of our star system. Later, it was realised that dust clouds obscured our view and hid most of the Galaxy from our sight.

By then, an even greater relegation of our place in the Universe had taken place – Hubble's proof that most of the *nebulae* were other galaxies. The Universe suddenly grew enormously larger than ever imagination could have predicted, but on Earth, life went on unaffected, and there was not even the slightest ripple on the stock exchanges.

We had suffered the greatest humiliation. The Sun, which had seemed to hold central spot, had been downgraded into insignificance. It was now just one of millions upon millions of stars in our Galaxy. It wasn't anywhere near the centre of the Galaxy. Worse still, far worse, our Galaxy was no longer the universe, but one amongst thousands, possible millions, of island universes.

The trend has continued. In the 20th century, each new telescope we built, each new detector we mounted, showed more and more galaxies. The large telescopes of today are capable of seeing perhaps a million, million galaxies. It leaves us as a speck of dust! Our planet, which once was held to dominate the Universe, has *in terms of its size* been reduced to total insignificance.

Yet, in one major respect, the Earth might be significant. It carries life.

We do not know all the forms that life could take, but we do know what it takes for life like us to exist. Water is the essential ingredient. Liquid, not ice, nor steam. To get this you need a planet with enough gravity to hold water as an ocean, and the planet has to orbit at the right distance from its parent star. Not too hot. Not too cold. Since we dwell on the land, and not within the oceans, continents are also a requisite.

Of course, the bulk of the planet is nevertheless not utilised. It may, like ours, have a radius of some six million metres, but the majority of the lifeforms dwell within *a layer of only a few metres thickness*. Excluding artificial means of

10.1 The Earth that was once thought to dominate the entire Universe has, in terms of size, been reduced to insignificance.

transport and elevation, we ourselves are confined to a layer a couple of metres deep stretching over the continents of the Earth. Our habitat is like a coat of paint put on part of a sphere – and a very thin coat of paint at that.

It seems astonishing, that in such a vast Universe, our habitat takes up so little room. The volume of the Earth is two million times larger than the thin layer we inhabit. The volume of the Solar System is two billion billion times larger still than the volume of the Earth. The figure is only approximate, since the Solar System has no definite boundary. If we presume that most other stars have families of planets, then there is a favourable chance that almost every Solar System has one planet neither too hot, nor too cold, but just right for liquid water to exist. If so, the layers suitable for habitation are but one part in a billion billion billion billion of the volume inside the Galaxy. Finally, if we take the spaces between the galaxies – and assume nobody is going to live there – then my calculation leads to the fact that *only one part of the Universe in nearly a hundred thousand billion billion billion billion is suitable for life!*

Given these incredible odds, how did we find the right place? Had we been placed at random anywhere in the Universe, it is extremely unlikely that it would have been fit for human habitation. Almost all of the Universe is unsuitable for life. How did we find the Earth?

The answer of course stares us in the face. Anywhere else, and we would not exist.

If we did not have a planet like Earth, we would not be here. If we did not have a Universe that produces Earth-like planets, we would not be here. Our very existence imposes constraints as to the nature of the Universe we inhabit. This relationship is known as the *Anthropic Principle*. No wonder the Universe looks like it does. It could hardly be otherwise, if we are here to see it.

Clearly, from the arguments above, we view our Universe from a very selected spot – an Earth-like planet that is just right for us. Could we also be seeing a very selected universe, one that is just right for us – one described by philosophers as a *Goldilock's* universe?

How is the Universe just right for us? For a start – because we live within an expanding Cosmic Egg. Thanks to the size of the egg, *we can only see a finite distance*. Thank goodness! Had we been able to see to greater distances, it could have been very uncomfortable. Suppose, for example, we choose one particular direction, and extend a line indefinitely out into space in that direction. If our line could keep going on almost forever and the Universe with it, then eventually it would impinge upon a star. That little bit of sky, to which the line is directed, would be as bright as the disk of that star. The disk would appear incredibly small, but its brightness would match the surface brightness of the Sun's disk. Extend similar lines over the entire sky, and by this argument, the entire sky would have the surface brightness of the Sun's disk. And we would be sizzled like sausages!

This argument is known as *Olber's Paradox*. It arises from the first thing that you notice about the night sky. *It is dark*. There may be stars, but they have plenty of dark sky between them. You cannot read a book by starlight. That immediately tells us that we cannot be looking an infinite distance into space.

But then an *infinite* distance would also mean an *infinite* age to the Universe, otherwise the light would not have had time to reach us. Rather, we know that the Universe has a *finite* age, and the *finite* distance that light has travelled within that finite age gives rise to the concept of the Cosmic Egg.

But the finite age came from a beginning, and that beginning was very hot – everywhere in the Universe. Surely, if we could see that hot big bang, we would be sizzled? Well, we cannot see the hot big bang itself, but we do see the early Universe when it was nearly as hot as the surface of the Sun – and we see it all over the sky. Still far too hot for comfort.

Saved again! This time by the *expansion* of the Universe. That expansion causes a *stretching*. The stretching robs the light of its energy, so much so that instead of seeing that early Universe at the 3,000 degrees that it was, it now appears at only 3 degrees. As we have seen, the radiation from the Cosmic Egg is in *microwaves*. The night sky is not dark, but our human eyes do not see it ablaze with microwave light.

We could not live in an infinite static Universe of infinite age, but we can survive in an expanding Universe of finite age.

There is another major reason why the Universe is just right for us. It lies in the laws of physics. Perhaps the best example (widely discussed in earlier chapters) is Newton's Law of Gravity. We are not about to get immersed in mathematics, except to point out that it involves a constant factor, known as *big G*. As any physicist will tell you, $G = 6.67 \times 10^{-11}$ Newtons metre^2 per kilogramme^2. But all we need to know is that it is 6.67 something or other.

Why 6.67? Why not 6.37 or 6.97? Surely it would make no difference? Yes it would. Tampering slightly with the value of *big G* would have enormous consequences for the lifetime of stars – and the lifetime of our Sun. If our Sun had a much shorter lifetime, then life would not have had time to evolve, or even develop, on Earth. In any case, it would not be Earth, as Earth and Mars would have been too hot, and only somewhere in the Asteroid Belt would have had the right temperature. If our Sun had a much longer lifetime, then its light would have been too feeble to provide the oasis that life needs.

Newton's Law of Gravity is but one example. There are plenty of other laws – and plenty of other constant factors, but all seem to have a value that is *just right* for tour existence. Indeed, if they did not, then we would not be here.

Are these laws of physics truly universal? Do they extend into a vastly greater Universe than just the Cosmic Egg? Even if they do, maybe they are confined only to *our Universe*. As physicists and philosophers speculate, it may be that our Universe is simply one of *a multitude of universes.*

Could there be other universes? It is a question beyond the bounds of science. By simple definition, other universes would not be accessible to us – unless some of that science fiction stuff is true. But as an armchair experiment, or philosophical consideration, we could imagine universes similar to our own, but where different conditions prevail.

Perhaps the laws of physics, or whatever governs their behaviour, are quite different in those other universes. You only have to change the laws of physics a little bit and a universe would look entirely different to the one we know. There could be hot universes, cold universes, mixed-up universes, and all sorts of universes. Out of the multitude came one just right for us – so it is not surprising that we find ourselves in it.

But this raises the inevitable question: How unique is *our* sort of life in *our* Universe?

Science may not have the answer, but it does have insight. Many scientists in the field are confident that primitive life might be abundant. The Earth's history strongly suggests it. For the first several hundred million years of its existence, the Earth would also have been inhospitable to life, due to the large number of disruptive impacts from meteors, as the debris left over from making the solar system was swept up. We know from the cratering on the Moon, that the impact rate thereafter rapidly declined. It seems that *as soon as it was safe for the Earth to carry life, so life did develop.*

That life, however, was primitive – most likely single-celled cyanobacteria, often described as blue-green algae. They are one of the most robust of lifeforms, even surviving in hot volcanic geysers. They are also the most enduring, for they still flourish today.

Of course, the key question – *how does such life spontaneously arise?* – has no clear answers. But science has been intrigued with an experiment first carried out in 1953 by Stanley Miller, who passed an electrical discharge through a chemically reducing mixture of gases. Such a mixture, rich in hydrogen compounds, likely formed the atmosphere of the early Earth. It is still present today as the atmospheres of the giant planets and the moon Titan (a large moon of Saturn). Lightning strikes provide natural electrical discharges. The outcome of Miller's experiment was the production of a mixture of organic compounds, including amino acids, the basis for the construction of proteins and life itself.

Continued experiments of this nature have produced the four complex molecules that form components for RNA and DNA, when they assemble, much as do fats and proteins, into long molecular chains. The RNA/DNA chains carry a particular sequence and are able to attract similar sets of molecules such that a corresponding sequence is built within a second chain. This is the principle for nature's photocopying machine – chains of molecules that can self-reproduce. It is the principle of life. Science can now generate some very simple self-reproducing molecules in the laboratory. They approach some of the primitive viruses, but are still a very very long way short of the living cell – the basis of the bacteria, algae and all larger lifeforms.

Where on Earth did life arise? The generally accepted notion that *warm tidal pools* would be suitable has, in recent years, been challenged by an alternative scenario that it may have come from *deep undersea volcanic vents.* Such vents, though remote and isolated in the blackness of the deep ocean, are nevertheless sources of warmth and nutrition. Today they are oases of life.

A still further alternative is that life might have come from space. We have recently realised that not only are planets pummelled over time by a great number of comets and meteors, but also they even interchange very small amounts of material. Impacts not only deposit but they blast small pieces out into space. Those pieces in time hit other planets. Of the meteorites found on Earth, most come from the Asteroid Belt, but a few originate from the Moon, and even from Mars. Though small rocks thrown off Mars would likely spend at least hundreds of thousands of years in space before they struck another body, it is considered possible that there might be simple lifeforms, in dormant form, that survive the journey.

Could life have even come from *other* solar systems? Are the seeds of life abundant in the Galaxy? The finding that interstellar clouds are rich in organic molecules boosted the old science-fiction idea that the seeds of life arrived from outer space. Does life arise spontaneously, or is seeding always necessary? Science does not yet have the answer.

In our Solar System, Mars, like Earth, is considered a potential place for life to have developed, because it used to have shallow oceans. Could life have first arisen on Mars, and then seeded the Earth? Mars today is cold and sterile. But in 1996, a claim that possible microfossils had been found in a Martian meteorite gave rise to enormous debate. Though the evidence now seems doubtful, that it was taken so seriously is significant.

Are there other planets in our Solar System that might give rise to life? Even the atmospheres of the giant outer planets offer possibility, since they certainly are sites of ongoing "Miller experiments" – though the first probe descending into Jupiter's atmosphere failed to reveal a rich array of organic molecules. Titan, the largest moon of Saturn, is considered a further possibility, and is accordingly the target of the *Huygens* mission, which is expected to drop a spacecraft into its cloudy atmosphere. Perhaps the best bet, however, is Europa, one of the large moons of Jupiter. It is surrounded by a thin crust of ice, beneath which is thought to be an ocean of water. Tidal stresses from Jupiter have warmed this moon's interior. It seems Europa is the only other body in the Solar System currently to possess abundant water in liquid form.

For most of its history, life on Earth has been nothing more than microscopic bacteria in the sea (see Figure 10.2). Life did not initially extend to the continents, as they received too much harsh ultra-violet radiation from the Sun. Had any human-like visitor arrived in the past, he or she would have probably concluded that the Earth was uninhabited.

Meteor or asteroid impacts may have declined to relatively safe levels – safe enough for life to have developed on Earth – but they have never ceased. A large impact can have global climatic consequences. Occasional large impacts have punctuated the Earth's history. They upset the balance of life. In more recent times, an asteroid (or large comet) impact most likely brought about the demise of the dinosaurs, yet thereafter allowed other life to take over. A similar impact (the Vredefort impact) is known to have occurred just over two billion years back, and it may be curiosity or coincidence that at that time, multicellular organisms seemed to develop for the first time.

Thereafter, little change seemed to occur, until only a half billion years ago, when there was an event as miraculous as the origin of life itself. This was the *Cambrian Explosion* – a sudden development of complex life. It followed

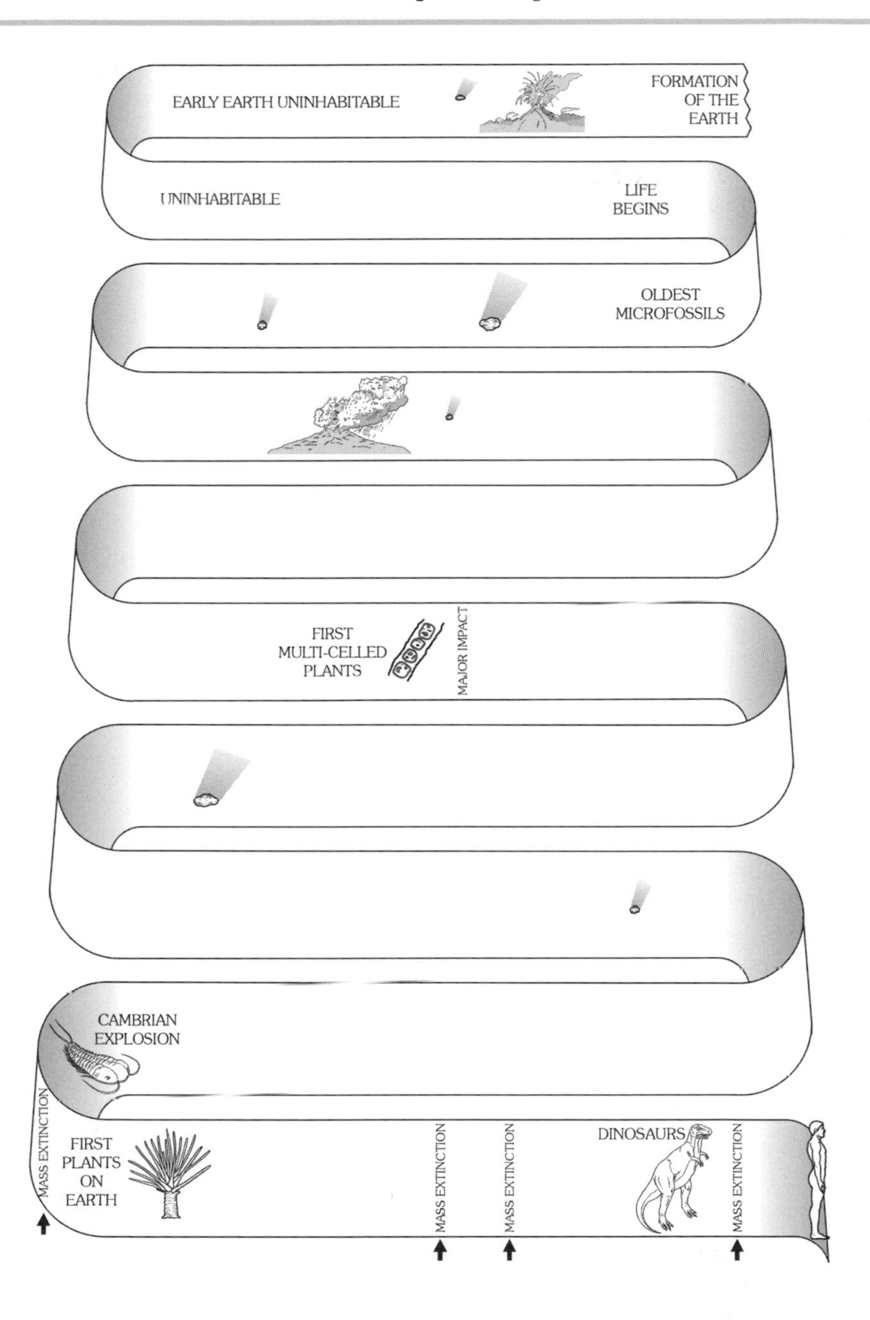

10.2 A timeline, of the development of life during the 4,500 million-year history of Earth, is depicted as a folded ribbon.

shortly after the first time "normal-sized" lifeforms – no longer microscopic – appeared on the planet. Cambrian life might be described as an array of "creepy-crawly" creatures. It was also the origin of the first true animals – that depended on eating plants or even other animals.

Almost all subsequent life on Earth traces to Cambrian roots; almost all plants and creatures alive today have their origins back in the Cambrian period. Fundamental developments then shaped the essential characteristics of life today. For instance, larger animals required skeletons for strength. Some animals developed hard outside skeletons. These are the arthropods – creatures with segmented bodies. They include not only the insect world, but crustaceans such as lobsters and crabs. They have multiplied into the greatest variety of living creatures. There are today more than a million different species of arthropods. Animals with skeletons on the inside also developed; the first vertebrates appeared some forty million years after the Cambrian Explosion. These creatures later evolved to form the basis for fish, reptiles and mammals.

Fossil records, however, show that most Cambrian life was suddenly extinguished (some 440 million years ago) – a likely indication that the planet suffered another devastating asteroid impact. What life survived multiplied to form a *different community* of creatures.

As life became larger and more abundant on Earth, so plant life absorbed more of the sunlight filtering into the uppermost layers of the ocean, and by the mechanism of photosynthesis, freed oxygen into the atmosphere. Oxygen allowed animals to metabolise much more efficiently and probably underpinned the development of large, hard-bodied creatures. The oxygen gradually formed a significant component in the air and an ozone layer high in the atmosphere. For the first time, the surface of the Earth was shielded from the Sun's ultraviolet rays, and for the first time it was safe for life – both plants and animals – to colonise the land as well as the sea.

But not entirely safe, as our records show that further mass extinctions occurred at 370 and 250 million years ago. Each time, the balance of life was upset and a new and different population of creatures subsequently established.

The most recent of the mass extinctions (that 65 million years ago) ended the long reign of the dinosaurs, though many of their relatives – creatures that hatch from eggs – have survived. Since then, we mammals have taken over and now dominate the land, and our species have grown physically larger as we flourished.

The largest creatures are, nevertheless, sometimes the most vulnerable. The giant dinosaurs were wiped out, but the small flying dinosaurs have survived to become the birds of today. In the last 50 million years, they have multiplied into some 10,000 species. But only 10,000 years ago, the Americas supported an enormous elephant population, which has since totally disappeared. We can be reasonably sure that it was not due to an asteroid impact, so obviously other factors – perhaps even biological – were at work. As we are all too uncomfortably aware, the increase and actions of humans today are reducing the numbers of large animals, and bringing about the extinction of numerous species.

We humans are the newcomers. Our species has only been around for a mere four or so million years, a *thousandth* of the Earth's history. It is almost as if we had just arrived on the planet. We are really strangers to our home. All that we might now call civilisation is but a few millionths of our planet's history. Our habitat *in time* is almost as selected as our habitat in space.

Yet, no matter these odds, many people – scientists included –are inclined to believe that there might currently be numerous other civilisations like ours, on other planets, and in other galaxies. In spite of the proliferation of "UFO" sightings during the period of the Cold War, and the spate of movies about aliens, it is obvious that our highways are not clogged with extraterrestrial craft; nor are our shopping malls full of visitors from space taking advantage of weekend specials. Whilst a small number of people may be believers, and UFOs sell well in tabloid journalism, the truth is that there is no *proof* that we have had any visits from an extraterrestrials civilisation. To most of us skeptical enquirers, there is no convincing evidence at all.

Until we explore our Solar System – and even other solar systems – in detail, we shall not know whether microscopic life can arise on other planets. We might suspect it is commonplace because the chemical building blocks of life, such as amino acids, appear to be able to form in interstellar space, because we find them in meteorites. But we could be completely wrong, and find that the Earth is the only place where any form of life exists.

But complex life and human-like civilisations are another matter. They may be so rare, that there are relatively few in the Galaxy. Or there may be none, even in the multitude of galaxies within the Cosmic Egg. Some investigators have invoked the *Strong Anthropic Principle* that our civilisation might be only one of its kind in the Universe. With only one known civilisation, who can say? But Science can be confident that there are many Earth-like planets available. Other investigators (this author included) would therefore argue that, in a Universe so huge, with perhaps at least a million, million galaxies, each of which has a million, million stars, it would seem impossible that we might be the only lifeforms to have reached this stage of intelligence (the so-called *Weak Anthropic Principle*).

Others would invoke a divine Creator for us being placed on this particular planet. This of course rests on belief, that it is simply the will of the Creator that has decided where in the Universe we should be. (Should that be the case, then Science could not read, nor attempt to understand, the mind of the Creator – and the matter would rest there.)

But, as we have seen, the triumph of Science is that it has found that the Universe is governed by a set of rules – the laws of physics. (If there is a Creator, then these laws are His rules.) The laws appear to apply throughout our observable Universe. They do not apply to one planet only, as was thought years ago. They do not apply to only one solar system, or to only one galaxy. Intelligent life on Earth must have evolved *according to those rules*. On that basis, the same could happen elsewhere. We may be a special civilisation in a special place, but our knowledge of the Universe strongly suggests that this special place is far from being unique.

Furthermore, our civilisation – which has so far only endured for a fleeting moment of cosmological time – might not encounter other civilisations because we do not happen to be around at the same time. As much as we might be separated by distance, so might we also be separated by time. Our perspective on our Universe is not only where we are, but *when* we are.

In a broader sense, the Anthropic Principle not only constrains the nature of the Universe (it must produce Earth-like planets), but it also constrains when we might dwell in the Universe.

We could not have existed in the very early Universe, the Universe seen in the "photograph" on the inside shell of the Cosmic Egg. It was then a Universe filled with incandescent gas. Galaxies, stars and planets had still to form.

Furthermore, there was nothing then with which to make us. Our form of life is often described as *carbon-based*. To go with the carbon, one needs oxygen, nitrogen, hydrogen and minor amounts of phosphorus, iron etc. But the intriguing thing is that the Big Bang produced only hydrogen and helium (and very small amounts of lithium). So the original Universe was *not* just right for us. But the *interiors of stars* have since processed the hydrogen as fuel, and some of the products of the nuclear reactions in stellar interiors were the heavier elements that we need for our bodies. Dying stars shed much of their matter back into interstellar space, where it would eventually accumulate into clouds massive enough to form a new generation of stars – and planets – and eventually, you and me.

We are, in a sense, children of the stars. Our existence is dependent on earlier stars generating the material from which we are made, and our presence is only possible because our planet orbits neither too close nor too far from a star, our Sun. So the Anthropic Principle not only requires a universe that produces Earth-like planets, but one that primarily produces stars – both to create heavy elements and later to warm the Earth-like planets.

Stars did not exist in the very early Universe. Probing to the earliest era of star formation is one of the main goals of the new generation of large telescopes, both those recently completed and some under construction (including those that operate outside normal optical wavelengths). Already, those telescopes have provided key information: the peak of star formation in the Universe is long since past. Star formation is currently on the decline.

Moreover, the star that gives us life, our Sun, will eventually bring about an end to life on Earth.

In a billion or two years time, the Earth will again become uninhabitable. As the hydrogen fuel reserves in the core of our Sun diminish, the nuclear "burning" that has so far been our survival will progress to a shell around the core. This will open up a fresh fuel supply, but the fuel will be consumed more readily. Consequently, the Sun will start to swell and its luminosity will increase. Gradually the Earth will get hotter. The oceans will in time evaporate, and the surface of our planet will be scorched. So life on Earth will end.

The Sun will, after this dramatic terminal period, shed its outer layers to interstellar space, while its core will likely condense to a compact white dwarf star, still incandescent, but a mere glowworm by comparison to its earlier glory.

So life for the Sun will end, as it must do for all stars. Stars run on fuel supplies, and once that fuel runs out in their cores, no more energy can be generated. Usually, there will still be abundant hydrogen supplies in the outer layers, and that fuel will probably be returned to the interstellar medium to enrich and to power a new generation of stars.

But eventually, eventually, the fuel will all be used up, and one by one the stars will go out. If the Universe survives into the distant future, tens or even hundreds of billions of years from now, then it will have only weakly glowing stellar remnants. But, like a dying fire, they too will be extinguished, and darkness will follow. It would be the end of the road of the Universe, as we know it. And, if it has not already happened, the end of life.

If, as some cosmologists have speculated, the clock of time runs relentlessly toward infinity, then even billions of years would become infinitely small intervals. The Universe that we currently know and love would be but a flash in the pan, a fleeting moment soon after its birth in an otherwise dark abyss.

The human race could not have survived in the distant past, nor can it survive into the remote future. We find ourselves living at just the right time and – like finding ourselves living in just the right place – it could not be otherwise, else we could not exist.

No wonder then that we find ourselves literally "at home in the Universe". In doing so, we have come full circle from the beginning we made in the opening chapter. Perhaps, as we move to a close, we might consider that in a cosmological sense, we are living at a very fortunate time.

Had this book been written a hundred years ago, its content would have been very different – the general consensus then was that our stellar system – our Galaxy – was the entire Universe. The recognition of other galaxies, the expansion of the Universe and Einstein's Relativity have all happened in the early 20th century. The Cosmic Egg – seen as the cosmic Microwave Background – was only detected in 1965. It is remarkable that only at this point in human history have we reached that final frontier. But it is the final frontier, and it would not be different even if this book was written in a hundred year's time. However, by then we should have far greater insight into the early Universe, thanks to the new telescopes now coming into operation.

It somehow seemed an appropriate analogy to describe that opaque spherical shell that appears to enclose our observable Universe as a Cosmic Egg. But an egg is a symbol of life; each of us began life as a single egg. So too is our cosmos a symbol of life. It has to be the way it is for our life to evolve and flourish.

Index